AF598045

METHODS IN MOLECULAR BIOLOGY™

Series Editor
John M. Walker
School of Life Sciences
University of Hertfordshire
Hatfield, Hertfordshire, AL10 9AB, UK

For further volumes:
http://www.springer.com/series/7651

Adenovirus

Methods and Protocols

Third Edition

Edited by

Miguel Chillón

Institució Catalana de Recerca i Estudis Avançats (ICREA) and CBATEG-Department of Biochemistry and Molecular Biology, Universitat Autonoma Barcelona, Bellaterra, Barcelona, Spain

Assumpció Bosch

Department of Biochemistry and Molecular Biology, Center of Animal Biotechnology and Gene Therapy (CBATEG), Universitat Autonoma Barcelona, Bellaterra, Barcelona, Spain

Editors
Miguel Chillón
Institució Catalana de Recerca i Estudis Avançats (ICREA) and CBATEG-Department of Biochemistry and Molecular Biology
Universitat Autonoma Barcelona
Bellaterra, Barcelona, Spain

Assumpció Bosch
Department of Biochemistry and Molecular Biology
Center of Animal Biotechnology and Gene Therapy (CBATEG)
Universitat Autonoma Barcelona
Bellaterra, Barcelona, Spain

ISSN 1064-3745 ISSN 1940-6029 (electronic)
ISBN 978-1-62703-678-8 ISBN 978-1-62703-679-5 (eBook)
DOI 10.1007/978-1-62703-679-5
Springer New York Heidelberg Dordrecht London

Library of Congress Control Number: 2013947923

Printed on acid-free paper

Humana Press is a brand of Springer
Springer is part of Springer Science+Business Media (www.springer.com)

Preface

Since the first reports in the mid-1950s, adenoviruses have been extensively studied in virology research, with over 43,000 publications listed in Pubmed (more than 1,700 in 2012 alone). Moreover, their efficiency as gene therapy tools has resulted in over 450 protocols approved for clinical trials. Initially, the majority of the techniques employed were commonly used in basic virology, such as small-scale production and purification, cell infection, and in vivo tropism. However, in recent years, advances in molecular biology, genomics and proteomics, imaging, and bioinformatics have allowed the development of powerful new tools and high-throughput methods to analyze adenoviral particles and their interactions with host cells, many of which are described in this book.

The *Adenovirus: Methods and Protocols* (3rd edition) book is intended for both the novice as well as the experienced investigators in the field. For this reason, despite it being impossible to cover all aspects of adenovirus research, we have aimed to compile a vast array of techniques as well as to review the most recent updates and new methodological developments in the field. Briefly, this third edition of *Adenovirus: Methods and Protocols* consists of 16 chapters covering state-of-the-art techniques. It includes cryoelectron microscopy, atomic force microscopy, and mass spectrometry for a high-resolution image and characterization of the virion (Chapters 1 and 2), capsid modifications and viral-like particles as promising alternatives to classical adenovirus vectors (Chapters 3 and 4), the study of adenovirus: host interactions in vitro at cellular level as well as in vivo in animal models (Chapters 5–10), and finally, an update of the most efficient protocols to generate, amplify, and/or purify at small- and large-scale, standard human Ad5 as well as nonhuman, chimeric, and helper-dependent adenovirus vectors (Chapters 11–16). More importantly, since this is a comprehensive methods book written by top scientists who are well aware of the pitfalls of the experiments, at the end of each chapter there is a helpful Notes section providing valuable troubleshooting guides and alternative procedures.

Adenovirus: Methods and Protocols (3rd edition) is the product of the hard work of a large number of scientists, all experts in various aspects of adenovirus research. We are indebted to all of them for their dedication and cooperation as well as for their patience during the editing process. We are also very grateful to the series editor, John Walker, as well as to David Casey and the editorial staff at Humana Press, for all the advice and assistance they have provided, and also to Rosemary Thwaite for her valuable assistance in proofreading. Without them, this book would not have been possible in its current form.

Finally, we would like to highlight that, as editors, it has been a privilege to participate in the development of the Adenovirus book. On behalf of all the authors, we hope you will find the book useful and informative.

Bellaterra, Barcelona, Spain ***Miguel Chillón***
Assumpció Bosch

Contents

Contributors

RAUL ALBA • *British Heart Foundation Glasgow Cardiovascular Research, Glasgow, UK; Nanotherapix S.L., Sant Cugat del Vallès, Spain*

RAMON ALEMANY • *Institut Catala d'Oncologia, Institut d'Investigació Biomèdica de Bellvitge, L'Hospitalet de Llobregat, Barcelona, Spain*

PAULA M. ALVES • *iBET, Instituto de Biologia Experimental e Tecnológica, Oeiras, Portugal; Instituto de Tecnologia Química e Biológica, Universidade Nova de Lisboa, Oeiras, Portugal*

KONSTANTIN A. ARTEMENKO • *Department of Chemistry-BMC, Uppsala University, Uppsala, Sweden*

ANDREW H. BAKER • *British Heart Foundation Glasgow Cardiovascular Research, Glasgow, UK*

ASSUMPCIÓ BOSCH • *Department of Biochemistry and Molecular Biology, Center of Animal Biotechnology and Gene Therapy (CBATEG), Universitat Autonoma Barcelona, Bellaterra, Barcelona, Spain*

ANGELA C. BRADSHAW • *British Heart Foundation Glasgow Cardiovascular Research, Glasgow, UK*

MIGUEL CHILLÓN • *Institut Català de Recerca i Estudis Avançats (ICREA) and CBATEG-Department of Biochemistry and Molecular Biology, Universitat Autonoma Barcelona, Bellaterra, Barcelona, Spain*

GABRIELA N. CONDEZO • *Department of Macromolecular Structure, Centro Nacional de Biotecnología (CNB-CSIC), Cantoblanco, Spain*

PEDRO J. DE PABLO • *Department of Physics of the Condensed Matter, Universidad Autónoma de Madrid, Madrid, Spain*

MARTA DEL ALAMO • *Department of Macromolecular Structure, Centro Nacional de Biotecnología (CNB-CSIC), Cantoblanco, Spain*

NELSON C. DI PAOLO • *Department of Medicine, Division of Medical Genetics, University of Washington, Seattle, WA, USA*

THOMAS DOBNER • *Department of Viral Transformation, Heinrich Pette Institute, Leibniz Institute for Experimental Virology, Hamburg, Germany*

PASCAL FENDER • *Unit of Virus Host Cell Interactions (UMI-3265:CNRS/UJF/EMBL), Grenoble, France*

PAULO FERNANDES • *Animal Cell Technology Unit, Instituto de Tecnologia Química e Biológica, Universidade Nova de Lisboa (ITQB-UNL), Oeiras, Portugal; Animal Cell Technology Unit, Instituto de Biologia Experimental e Tecnologica (IBET), Oeiras, Portugal*

MARC GARCIA • *Department of Biochemistry and Molecular Biology, Center of Animal Biotechnology and Gene Therapy (CBATEG), Universitat Autonoma Barcelona, Bellaterra, Spain*

RAUL GIL-HOYOS • *Institut Catala d'Oncologia, Institut d'Investigació Biomèdica de Bellvitge, L'Hospitalet de Llobregat, Barcelona, Spain*

PATRICK HEARING • *Department of Molecular Genetics and Microbiology, School of Medicine, Stony Brook University, Stony Brook, NY, USA*

FLORIAN KREPPEL • *Department of Gene Therapy, University of Ulm, Ulm, Germany*
FREDERIC LEMNITZER • *Max von Pettenkofer Institute, Ludwig-Maximilians-University, Munich, Germany*
SARA BERGSTRÖM LIND • *Department of Immunology, Genetics, and Pathology, Rudbeck Laboratory, Uppsala University, Uppsala, Sweden*
ROSA MENÉNDEZ-CONEJERO • *Department of Macromolecular Structure, Centro Nacional de Biotecnología (CNB-CSIC), Cantoblanco, Spain*
JUAN MIGUEL-CAMACHO • *Institut Catala d'Oncologia, Institut d'Investigació Biomèdica de Bellvitge, L'Hospitalet de Llobregat, Barcelona, Spain*
MARTA MIRALLES • *Department of Biochemistry and Molecular Biology, Center of Animal Biotechnology and Gene Therapy (CBATEG), Universitat Autonoma Barcelona, Bellaterra, Spain*
SUSANA MIRAVET • *Unitat Producció Vectors (UPV), CBATEG, Department of Biochemistry and Molecular Biology, Universitat Autonoma Barcelona, Bellaterra, Spain*
MERCÈ MONFAR • *Unitat Producció Vectors (UPV), CBATEG, Department of Biochemistry and Molecular Biology, Universitat Autonoma Barcelona, Bellaterra, Spain*
MICHAEL NEVELS • *Institute for Microbiology and Hygiene, University of Regensburg, Regensburg, Germany*
STUART A. NICKLIN • *British Heart Foundation Glasgow Cardiovascular Research, Glasgow, UK*
MARIA ONTIVEROS • *Unitat Producció Vectors (UPV), CBATEG, Department of Biochemistry and Molecular Biology, Universitat Autonoma Barcelona, Bellaterra, Spain*
ALVARO ORTEGA-ESTEBAN • *Department of Macromolecular Structure, Centro Nacional de Biotecnología (CNB-CSIC), Cantoblanco, Spain; Department of Physics of the Condensed Matter, Universidad Autónoma de Madrid, Madrid, Spain*
ALAN L. PARKER • *Institute of Cancer and Genetics, Cardiff University, Cardiff, UK*
CRISTINA PENALVA • *Department of Biochemistry and Molecular Biology, Unitat Producció Vectors (UPV), CBATEG, Universitat Autonoma Barcelona, Bellaterra, Spain*
ANA J. PÉREZ-BERNÁ • *Department of Macromolecular Structure, Centro Nacional de Biotecnología (CNB-CSIC), Cantoblanco, Spain*
ULF PETTERSSON • *Department of Immunology, Genetics, and Pathology, Rudbeck Laboratory, Uppsala University, Uppsala, Sweden*
JOSE PIEDRA • *Department of Biochemistry and Molecular Biology, Unitat Producció Vectors (UPV), CBATEG, Universitat Autonoma Barcelona, Bellaterra, Spain*
STEFANIA PIERSANTI • *Dipartimento di Biologia e Biotecnologie, Sapienza, Università di Roma, Roma, Italy*
MERITXELL PUIG • *Department of Biochemistry and Molecular Biology, Center of Animal Biotechnology and Gene Therapy (CBATEG), Universitat Autonoma Barcelona, Bellaterra, Spain*
ZSOLT RUZSICS • *Max von Pettenkofer Institute, Ludwig-Maximilians-University, Munich, Germany*
ISABELLA SAGGIO • *Dipartimento di Biologia e Biotecnologie, Sapienza, Università di Roma, Roma, Italy; IBPM, CNR, and Fondazione Cenci Bolognetti, Bolognetti, Italy*
SARA SALINAS • *Institut de Génétique Moléculaire de Montpellier, CNRS UMR 5535, Universités de Montpellier I & II, Montpellier, France*
CARMEN SAN MARTÍN • *Department of Macromolecular Structure, Centro Nacional de Biotecnología (CNB-CSIC), Cantoblanco, Spain*

MARÍA MERCEDES SEGURA • *Departamento d`Enginyeria Química, Center of Animal Biotechnology and Gene Therapy (CBATEG), Universitat Autònoma de Barcelona, Bellaterra, Spain*
DMITRY M. SHAYAKHMETOV • *Department of Medicine, Division of Medical Genetics, University of Washington, Seattle, WA, USA*
ANA CARINA SILVA • *Animal Cell Technology Unit, Instituto de Tecnologia Química e Biológica, Universidade Nova de Lisboa (ITQB-UNL), Oeiras, Portugal; Animal Cell Technology Unit, Instituto de Biologia Experimental e Tecnologica (IBET), Oeiras, Portugal*
MARCOS F.Q. SOUSA • *Animal Cell Technology Unit, Instituto de Tecnologia Química e Biológica, Universidade Nova de Lisboa (ITQB-UNL), Oeiras, Portugal; Animal Cell Technology Unit, Instituto de Biologia Experimental e Tecnologica (IBET), Oeiras, Portugal*
THOMAS SPEISEDER • *Department of Viral Transformation, Heinrich Pette Institute, Leibniz Institute for Experimental Virology, Hamburg, Germany*
ENRICO TAGLIAFICO • *Dipartimento di Scienze biomediche, Università di Modena, Modena, Italy*
CHRISTIAN THIRION • *Sirion Biotech GmbH, Martinsried, Germany*
MARCOS TEJERO • *Department of Biochemistry and Molecular Biology, Center of Animal Biotechnology and Gene Therapy (CBATEG), Universitat Autonoma Barcelona, Bellaterra, Spain*
YUETING ZHENG • *Department of Molecular Genetics and Microbiology, School of Medicine, Stony Brook University, Stony Brook, NY, USA*
CHARLEINE ZUSSY • *Institut de Génétique Moléculaire de Montpellier, CNRS UMR 5535, Universités de Montpellier I & II, Montpellier, France*

Chapter 1

Biophysical Methods to Monitor Structural Aspects of the Adenovirus Infectious Cycle

Rosa Menéndez-Conejero, Ana J. Pérez-Berná, Gabriela N. Condezo, Alvaro Ortega-Esteban, Marta del Alamo, Pedro J. de Pablo, and Carmen San Martín

Abstract

In this chapter we compile a battery of biophysical and imaging methods suitable to investigate adenovirus structural stability, structure, and assembly. Some are standard methods with a long history of use in virology, such as embedding and sectioning of infected cells, negative staining, or immunoelectron microscopy, as well as extrinsic fluorescence. The newer cryo-electron microscopy technique, which combined with advanced image processing tools has recently yielded an atomic resolution picture of the complete virion, is also described. Finally, we detail the procedure for imaging and interacting with single adenovirus virions using the atomic force microscope in liquid conditions. We provide examples of the kind of data obtained with each technique.

Key words Adenovirus structure, Adenovirus assembly, Adenovirus stability, Atomic force microscopy, Fluorescence spectroscopy, Electron microscopy, Immunolabeling, Ultrathin sections

1 Introduction

The quality of adenoviral preparations is usually judged from their infectious titer (plaque assay or CPE detection), genome analysis (restriction pattern or PCR), or transduction of visible markers (e.g., GFP). Structural and biophysical techniques can provide additional information regarding virion structure, assembly, and stability, and can help determine the ultimate causes of compromised infectivity in new constructs or preparations. Here we gather together a collection of biophysical and imaging procedures that represent alternative, complementary methods to evaluate the structural integrity of an adenovirus preparation, or to elucidate whether the viral cycle is impaired at a particular stage in the cell.

We begin with the determination of viral stability using *fluorescence spectroscopy*. Determination of the structural stability of a viral

Miguel Chillón and Assumpció Bosch (eds.), *Adenovirus: Methods and Protocols*, Methods in Molecular Biology, vol. 1089, DOI 10.1007/978-1-62703-679-5_1, © Springer Science+Business Media, LLC 2014

stock under a variety of conditions is important not only for laboratory work, but particularly for translational purposes that require carefully controlled storage conditions. Fluorescence is defined as the emission of light by a substance that has absorbed electromagnetic radiation. Proteins contain three types of aromatic amino acid residues (tryptophan, tyrosine, and phenylalanine) that contribute to their intrinsic fluorescence. These residues produce different fluorescence emission (both in wavelength and intensity) depending on the polarity of the environment, with the final result that the intrinsic fluorescence of proteins varies with their folding state. Therefore, measurements of intrinsic fluorescence intensity serve to probe perturbations in folding. However, in the case of a large macromolecular complex such as adenovirus, intrinsic fluorescence measurements are difficult to interpret. An alternative methodology is the addition of dyes whose fluorescent properties reflect particular characteristics of the sample. This technique is known as extrinsic fluorescence. Measurement of extrinsic fluorescence using propidium iodide (PI) as a dye can be used to determine adenovirus stability in a straightforward way. When intact, the viral capsid is impermeable to PI. When the viral preparation is subject to increasing stress levels (temperature, pH, ionic strength, etc.), at a certain point the threshold for capsid structural integrity will be exceeded and the capsid breaks, allowing access of PI to the genome inside. PI binds to dsDNA by intercalating between the bases, which causes a 20–30-fold enhancement of its fluorescent emission. Therefore, monitoring PI fluorescence emission as stress increases reveals the condition at which the capsid opens and exposes the viral genome to the environment [1, 2].

The rest of the chapter is devoted to visualization techniques. Three sections deal with different flavors of electron microscopy (EM) aimed to image purified virus preparations; we then introduce the use of atomic force microscopy (AFM) for imaging and interacting with single virions without drying, in buffer conditions; and finally, we describe two alternative techniques to prepare infected cells for visualization at the electron microscope. Because of the expensive equipment required, and the constant evolution of the field, use of these techniques usually requires either access to a state-of-the-art infrastructure facility, or collaboration with an expert group.

The simplest preparation technique to examine virus morphology by EM is *negative staining* [3]. Negative staining agents are compounds of metals with high atomic number that serve to scatter the electrons from regions covered with the stain, therefore enhancing the contrast between the background and the biological material in the image. The stain is excluded from the virus particle and forms a cast around it, outlining its structure. It is also able to penetrate between small surface projections to delineate them. Areas where stain accumulates are dark in the image, while the excluded areas (containing the biological material) are light—therefore the "negative" in the technique's name. Negative staining is a fast technique that requires small sample amounts. In a large and faceted icosahedral particle like the adenovirus virion, individual capsomers can be discerned using this technique. In fact, adenovirus was one of the first viruses to be imaged at the electron microscope [4].

The purely morphological information provided by negative staining can be complemented with labeling techniques to detect specific components in the sample. Labeling is particularly relevant when imaging a sample with complex protein composition such as adenovirus. The *negative staining immunoelectron microscopy* technique [5] can be used to detect the presence, location, and accessibility of proteins or peptides in the virion. In this technique, the virus is adsorbed onto a carbon support film, followed by a series of "on grid" labeling and washing steps and a final negative staining. The bound antibody is visualized using a secondary antibody conjugated with gold particles. After negative staining, the gold particles show up in the microscope as black dots.

Negative staining is a straightforward, extremely valuable sample preparation method for EM visualization of viruses. However, it must always be considered that the specimen is observed in conditions quite different from its physiological state: staining agents may have extreme pH values; the sample is dried; and the image represents the cast left by the embedding heavy metal, while the actual biological material is destroyed during imaging due to the high vacuum and radiation conditions in the microscope. *Cryo-electron microscopy* (cryo-EM) allows visualization of samples in a more close-to-native environment [6]. For cryo-EM, a thin layer (~1,000 Å) of sample preparation in an appropriate buffer solution is frozen by immersion in a cryogen (e.g., liquid ethane), at a speed high enough to prevent formation of ice crystals that would destroy the specimen structure [7]. The sample, in this frozen hydrated state, is imaged at low temperatures (ca. −180 °C) [8]. Unlike negative staining, cryo-EM images contain information directly related to the specimen density at each point, including the interior of viral particles. A disadvantage of cryo-EM is that images have very low signal to noise ratio, due to the lack of contrasting agent and to the use of very low electron doses in image acquisition to prevent radiation damage. As a consequence, cryo-EM images are often hard to interpret in a direct way. Both negative staining and cryo-EM provide projection images, merging in a single plane the information coming from the different heights in the object. Three-dimensional (3D) information can be recovered using sophisticated image processing methods [9, 10]. Adenovirus was also one of the first specimens to be studied in 3D using cryo-EM [11]. Technical improvements in the past years resulted in the recent solution of the virion structure at 3.5 Å resolution, similar to that obtained by protein crystallography [12, 13]. Cryo-EM studies have also revealed details on adenovirus receptor or antibody binding (*reviewed in* [14]), as well as uncoating [2].

The most novel technique we address here is *Atomic Force Microscopy* (AFM) [15]. AFM was first described in 1986, and has been used in different research fields since then, most notably in materials science. The atomic force microscope belongs to the scanning probe microscope (SPM) family. It consists of a sharp tip at the end of a cantilever, which bends depending on the interactions between the tip and the sample. The tip sharpness and the scanning conditions play an important role in the AFM resolution, which under the appropriate conditions can even reach atomic level [16]. AFM has become a widely used tool in biophysics, due to its ability to apply forces as small as piconewtons (pN) and resolve structures in the μm to nm range [17]. AFM provides information on the topography of the sample surfaces that are accessible to the tip, but not on internal features. Apart from providing this purely visual information, the specific potential of AFM lies in its ability to interact with the specimen and measure forces at single molecule resolution. AFM allows to measure physical properties such as stiffness or surface charge density, among others [18–20]. In physical virology, AFM is helping to relate structure to physical properties in single viral particles [2, 21–23].

All the techniques mentioned so far refer to analyses of purified virus particles. When *visualization of infected cells* at the electron microscope is desired, specific preparation techniques must be used to allow structural preservation, sectioning at the required thickness (~70 nm) and contrast generation. Different variations of chemical cross-linking, resin embedding, and staining can be used depending on the particular experimental goal. Here we describe a standard procedure for room temperature embedding, aimed at providing high contrast and structural preservation for morphological characterization. We also indicate a variant of the technique consisting in a milder, low temperature embedding procedure, providing conditions in which immunogenic properties are preserved to allow labeling on cell sections.

2 Materials

2.1 Fluorescence Spectroscopy

1. Sealed quartz cuvette with a light path of 10 mm, and volume of 700 μL (Hellma® absorption cuvette or similar).
2. Spectrofluorimeter equipped with filters for excitation at 353 nm and emission monitoring from 550 nm to 650 nm and a Peltier cooler (Hitachi Model F-2500 FL or equivalent).
3. Purified adenovirus sample at 5×10^{10} viral particles/mL (vp/mL) concentration.

4. Appropriate buffer without dye traces (PBS, Tris–HCl, HEPES, etc.).
5. Propidium iodide. To prepare a stock solution from the solid form, dissolve PI (MW = 668.4) in deionized water or buffer at 1 mg/mL (1.5 mM) and store at 2–6 °C, protected from light. Caution: PI is a potential mutagen and should be handled with care. It must be disposed of safely and in accordance with applicable local regulations.
6. Origin software for data analysis (http://www.originlab.com).

2.2 Negative Staining Transmission Electron Microscopy

1. Copper grids covered with carbon or carbon/collodion film (for example GILDER GRIDS G400-C3, 400 lines/in. Square Mesh Copper).
2. Glow discharge equipment (Emitech K100x or equivalent).
3. High precision grade tweezers (high precision grade Dumont Tweezers type 4, stainless steel anti-mag or equivalent).
4. Whatman Grade N°1 Filter paper, 90 mm diameter.
5. Purified adenovirus sample.
6. Staining agent: uranyl acetate. Usually prepared as a 1–2 % (w/v) solution in double-distilled water without adjustment of the initial pH of approximate pH 4.3–4.5 (it precipitates above pH 5.0). Uranyl acetate solutions must be protected from light and filtered using 0.025 μm filters prior to use. Caution: uranyl acetate has heavy metal solution waste category and requires following local regulations for heavy metal handling, safety, and disposal. It is toxic by ingestion and if inhaled as dust. Use of gloves and safety goggles is required when preparing the diluted reagent.
7. Access to a 100/120 kV transmission electron microscope (JEOL-JEM 1100 or similar).

2.3 Negative Staining Immunoelectron Microscopy of Purified Viruses

1. **Items 2–7** in Subheading 2.2.
2. Carbon or collodion/carbon coated nickel EM grids (*see* **Note 1**).
3. Tris-buffer saline (TBS: 10 mM TRIS-base, HCl up to pH 8.2, 150 mM NaCl).
4. Bovine serum albumin (BSA).
5. Cold water fish skin gelatin.
6. Blocking solutions: TBG: TBS + 0.1 % BSA + 1 % gelatin. TBS/BSA: TBS + 1 % BSA.
7. Purified adenovirus sample.
8. Primary antibody of interest, as well as positive and negative control antibodies. For example, an antibody against hexon can be used as a positive control, and an irrelevant IgG as a negative control.
9. Secondary antibody conjugated to 10 nm colloidal gold particles (e.g., Goat Anti-mouse IgG + IgM EM-grade 10 nm cat# 25168 Electron Microscopy Sciences) (*see* **Note 2**).

2.4 Cryo-Electron Microscopy

1. Holey carbon covered grids (e.g., Quantifoil R 2/4).
2. Manual or automated plunge-freezing device (e.g., Leica CPC, FEI Vitrobot) [7].
3. High precision grade tweezers compatible with plunge-freezing device to be used (e.g., Dumont #L5 Medical Forceps with Slide for Leica CPC).
4. Liquid nitrogen. Caution: extremely cold. Wear safety eyewear. Use in well-ventilated areas.
5. Ethane gas. Caution: highly flammable. Do not use in the presence of an open flame. Wear safety eyewear. Dispose by allowing to evaporate in fume hood. Use in well-ventilated areas.
6. Whatman Grade N° 1 Filter paper, 90 mm diameter.
7. Purified virus sample at 5×10^{12} vp/mL in sucrose or glycerol free buffer (*see* **Note 3**).
8. Cryo-grid boxes for storage and transfer under liquid nitrogen (e.g., Ted Pella cat# 160-40).
9. Liquid nitrogen tank for storage of vitrified samples (AirLiquide GT35 or similar).
10. Access to a transmission electron microscope (FEI Tecnai G20 or similar) equipped with cryo-holder (e.g., GATAN 626) and low dose image acquisition system (*see* **Note 4**).

2.5 Atomic Force Microscopy

1. Atomic force microscope equipped with operating and image processing software WSxM [24] (e.g., Nanotec Dulcinea, Nanotec, Madrid, Spain).
2. Standard Silicon Nitride Rectangular RC800 (20×200 μm, 0.05 N/m and 40×200 μm, 0.1 N/m) and Biolever RC150 (30×60 μm, 0.03 N/m) cantilevers (Olympus, Japan).
3. Purified adenovirus sample at a concentration of ~5×10^{12} vp/mL stored in HBS buffer (20 mM HEPES, 150 mM NaCl, pH 7.8) in single-use (5 μl) aliquots at −70 °C.
4. Stock solution of $NiCl_2$ in HBS buffer at the required concentration to give 5 mM Ni^{2+} in the final sample.
5. Flat, clean surface for sample attachment and support (e.g., Muscovite mica V-1 quality) (*see* **Note 5**).

2.6 Electron Microscopy of Infected Cells

2.6.1 Fixation, Dehydration, and Embedding in Epoxy Resins for Morphological Studies

1. Adenovirus-infected and mock-infected cells at the appropriate post-infection time(s), grown in monolayer in 10 cm diameter culture plates. Usually one plate is enough to obtain three blocks of sample for sectioning.
2. 0.4 M HEPES buffer, pH 7.2.
3. PBS: 8.01 g/L NaCl, 0.20 g/L KCl, 1.78 g/L $Na_2HPO_4 \cdot 2H_2O$, 0.27 g/L KH_2PO_4, pH 7.4.

4. EM-grade 25 % glutaraldehyde. After opening the vial, store at 4 °C. Glutaraldehyde can irritate the skin and mucous membrane by direct contact or vapor inhalation. It must always be handled in a fume hood.
5. Tannic acid.
6. *En bloc* fixative/staining agent: 1 % osmium tetroxide (OsO_4), 0.8 % potassium ferricyanide ($K_3[Fe(CN)_6]$) in water. Prepare the mixture on the same day and protect from light. Caution: OsO_4 must always be handled in a fume hood and skin contact must be avoided at all times.
7. *En bloc* staining agent: uranyl acetate 2 % w/v (protect from light). Follow safety measures as described in previous sections.
8. Dehydration agent: dried acetone.
9. Embedding medium: 812 Epon Embedding Kit (Electron Microscopy Sciences). Caution: some of the kit components are irritating to the eyes and respiratory tract; others are corrosive and flammable. Do not get into eyes, on skin, or clothing. Use only with adequate ventilation. Do not breathe vapor. Follow the MSDS instructions for each kit component.

   For 10 mL of Epon, weigh:

   (a) 812 resin: 4.8 g.

   (b) DDSA (Dodecenyl Succinic Anhydride): 1.9 g.

   (c) MNA (Methyl Nadic Anhydride): 3.3 g.

   (d) Accelerator BDMA (Benzyl Dimethylamide): 5 drops (from a 3 mL plastic Pasteur pipette).

   All these substances are dense liquids. To weigh them, use a 3 mL plastic Pasteur pipette to add gradually the resin to a Falcon tube placed on the scale. Mix well. Centrifuge the mixture before using it to remove air bubbles (15,900 × *g* for 10 min).
10. Wooden toothpicks.
11. Encapsulation: Beem capsules size 3 (Electron Microscopy Sciences).

2.6.2 Sectioning

1. Diamond knife (Diatome 45°).
2. Ultramicrotome (Leica EM UC7 or similar).
3. Single edge blades (TAAB cat#B056 or equivalent).
4. Wire loop (Diatome Perfect Loop, Electron Microscopy Sciences).
5. Glass knife maker (e.g., RMC Products GKM-2 or similar).
6. Glass knife strips (Electron Microscopy Sciences).
7. Needles (1.2 × 40 mm 18G).

8. Glass microscope slides.
9. 1 % toluidine blue, 1 % sodium tetraborate in water.
10. Single hair attached to a wooden handle.
11. 200 mesh nickel EM grids (Veco), coated with 0.25 % Formvar in chloroform.

2.6.3 Section Staining and Observation

1. Saturated uranyl acetate (8–10 % w/v in water). Protect from light. Follow safety measures as described in previous sections.
2. 0.2 % lead citrate in 0.4 % NaOH. Store at 4 °C and seal with Parafilm to avoid formation of precipitates due to contact with oxygen. Caution: avoid contact with eyes, skin and clothing. Avoid ingestion and inhalation.
3. Whatman filter paper.
4. Access to a 100/120 kV transmission electron microscope (JEOL-JEM 1100 or similar).

3 Methods

3.1 Fluorescence Spectroscopy

1. Prepare a 100 μL sample of completely disrupted adenovirus by heating it at 65 °C for 15 min (*see* **Note 6**).
2. Mix PI into the sample and take an emission spectrum from 580 to 700 nm, exciting at 535 nm. Note the position and intensity of the emission maximum.
3. Repeat **steps 1–2** with different virus and PI concentrations to determine the conditions producing maximum fluorescence emission. The PI fluorescence maximum when bound to dsDNA should occur at approx. 617 nm emission wavelength.
4. Adjust the excitation and emission slit widths and spectrofluorimeter recording conditions to avoid signal saturation. This will be the maximum signal to be expected for the experiment.
5. Take an emission spectrum in the same conditions for a sample containing only buffer and PI at the concentration determined in **step 3** (blank). This will be the basal data to use for later subtraction. Note that different blank spectra may need to be obtained if the experiment involves the use of different buffers (e.g., if testing stability against high ionic strength, or against acidification).
6. Take an emission spectrum of the adenovirus sample in physiological conditions, using the virus and PI concentrations determined in **step 3**. Record emission from 580 to 700 nm, exciting at 535 nm. Subtract the blank emission spectrum taken in **step 5**. The intensity value of the corrected spec-

trum at the emission maximum is I_0, initial fluorescence of intact virus.

7. Take a new virus sample subject to a particular stress condition. For example, to investigate thermostability, heat independent samples at different temperatures for 15 min, add PI to the concentration determined in **step 3**, equilibrate for 5 min at the same temperature in the fluorimeter cell, and take their emission spectra as described in **step 6** (*see* Fig. 1a).
8. Correct each spectrum by subtraction of the correct blank intensities for each condition.
9. Make the experiment in triplicate by recording three independent spectra for each condition tested. Calculate the maximum emission intensity +/− the standard errors for the three independent spectra for each condition, after subtraction of the corresponding blank spectrum.
10. Plot the ratio of PI fluorescence to the initial emission (I/I_0) at the maximum position as a function of stress condition values (*see* Fig. 1b).

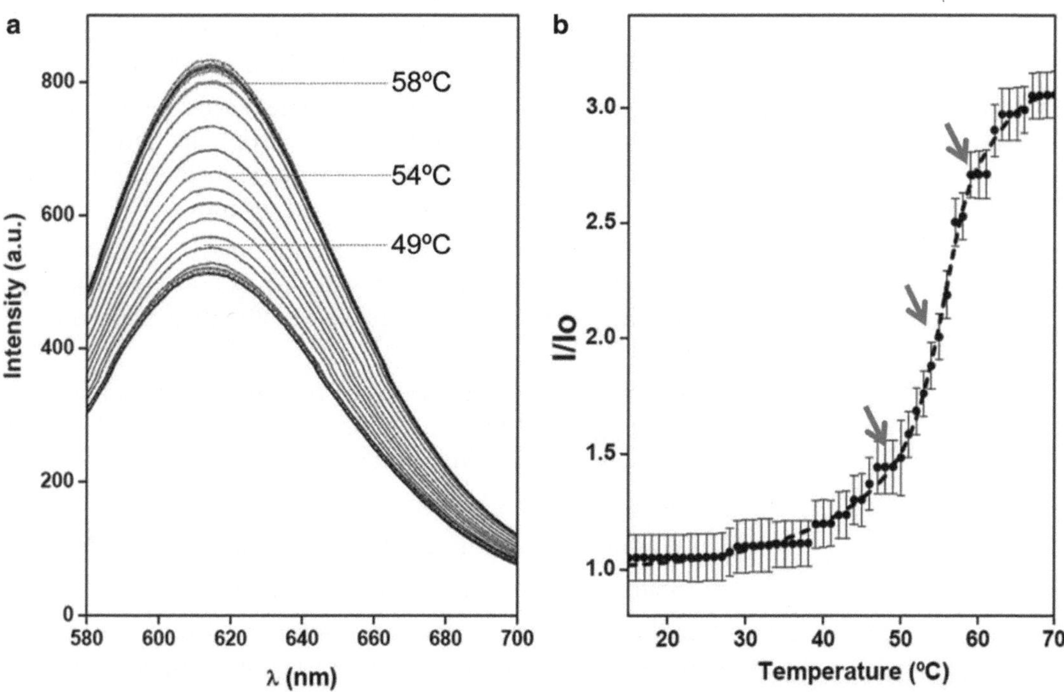

Fig. 1 Adenovirus thermal stability studied by PI fluorescence spectroscopy. (**a**) Series of raw emission spectra taken at different temperatures. Some temperature values are indicated. (**b**) Extrinsic fluorescence curve obtained after data processing and experiment replication (error bars for three independent measurements per temperature). *Arrows* indicate the temperature points showed in (**a**). Data correspond to immature virions produced by the Ad2 *ts1* mutant [2]

11. Fluorescence intensity changes can be fitted to a sum of sigmoids using Origin, according to the following expression:

$$I / I_0 = A_1 + \frac{B_1 - A_1}{1 + \exp\left(\frac{C_1 - x}{C_1 D_1}\right)} + \sum_{i=2}^{n} \left[B_{i-1} + \frac{B_i - B_{i-1}}{1 + \exp\left(\frac{C_{i-1} - x}{C_i D_i}\right)} \right]$$

where A_i and B_i are the lower and upper platform values for each sigmoid (notice that starting from the second sigmoid, the lower platform A_i is forced to match with the previous upper platform B_{i-1}); C_i is the transition midpoint; and D_i is the slope. The number of sigmoid curves in the summatory (up to $n=3$) is taken as the minimum necessary to fit the experimental curves, based on best R^2 values. Higher slope values indicate high cooperativity effects.

3.2 Negative Staining Transmission Electron Microscopy

1. Prepare several dilutions of the purified adenovirus sample. A virus concentration around 1×10^{11} vp/mL usually gives good results, but since the spreading of particles on the grid may vary depending on the hydrophobicity of the carbon support, it is generally a good idea to try different dilutions every time, even if using samples with the same concentration.
2. Glow discharge collodion/carbon coated grids. Glow discharge makes the carbon surface more hydrophilic and facilitates homogeneous spreading of the particles on the support. Use glow discharged grids within the next 30 min.
3. Place a piece of Parafilm (about 20 cm long) on the bottom of a large petri dish and apply pressure to its corners to attach it to the dish bottom. Place a piece of Whatman paper next to the dish.
4. Prepare three lines with drops of the samples (first line) (one drop for each dilution), washing buffer (second line), and staining agent (third line) on the Parafilm strip as follows:
   (a) Sample: one 5 μL drop.
   (b) Washing buffer: three 50 μL drops (filtered) (*see* **Note 7**).
   (c) Staining agent: one 10 μL drop (filtered).
5. Place the grid over the first drop (sample) with the carbon side facing the drop. Wait for 3–5 min to allow sample adsorption.
6. Blotting: Use high precision tweezers to hold the grid at the continuous ring of metal at the edge. Carefully, remove the excess fluid by touching on the edge of the grid with a piece of Whatman paper, without allowing it to become completely dry.

7. Washing: Place the grid (carbon side down) on the first washing buffer drop, remove and blot immediately as above. Repeat the washing/blotting step two more times.
8. Staining: Place the grid on the staining agent drop. Wait for 30 s to 1 min.
9. Pick up the grid, blot, and leave it (facing up) to air-dry on filter paper in a covered petri dish.
10. Once dried (1–3 h), the grid is ready for observation at the electron microscope (*see* Fig. 2a, b) (*see* **Note 8**).

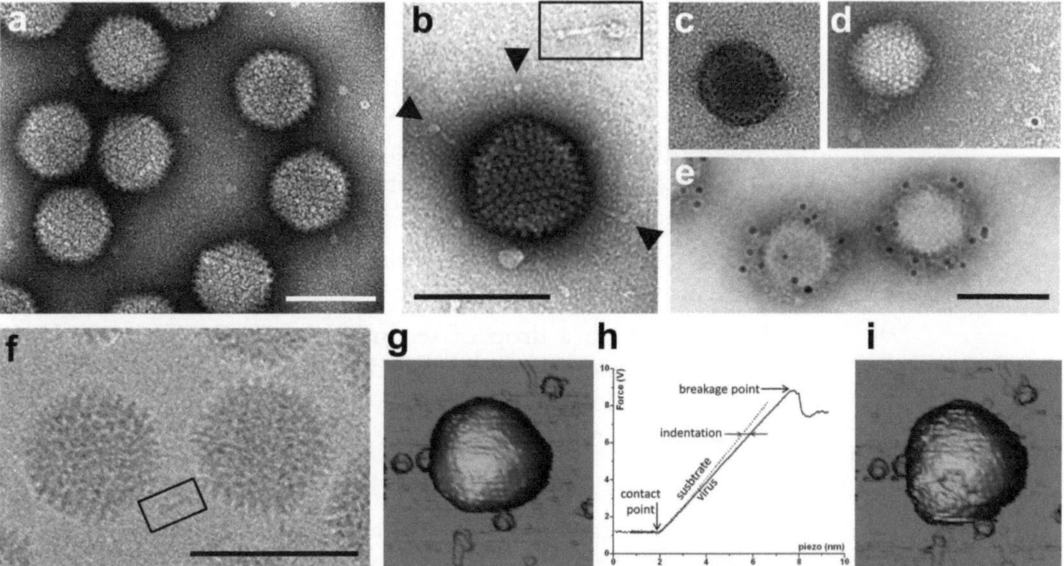

Fig. 2 Various techniques to image and characterize purified adenovirus virions. (**a**) General view of a negatively stained electron microscopy preparation. Capsomers appear white, while accumulation of staining agent around the large viral particle results in a black halo surrounding the virion. (**b**) Detail of a negatively stained virion where attached fibers are visible (*arrowheads*). The *black rectangle* indicates a loose penton complex (penton base plus fiber) in the background. Since the height of fibers is much smaller than the virion height on the substrate, no prominent black halo is observed around them. (**c**) Example of a positively stained viral particle. Notice that in this kind of image structural detail is lost. Grid areas presenting this kind of stain should be avoided. (**d, e**) Negative staining immunoelectron microscopy. Ad5-derived vectors are probed with an anti-FLAG monoclonal antibody followed by a goat anti-mouse IgG conjugated with 5 nm colloidal gold particles. In (**d**), the negative control with Ad5GL vector (structurally wild type) is shown. In (**e**), the antibodies have detected the presence of the FLAG peptide fused to the C-terminus of polypeptide IX [38]. (**f**) Adenovirus imaged by cryo-EM. The contrast is the opposite as that of negative staining. A loose penton complex is indicated by a *black rectangle*. (**g**) Example of one adenovirus virion imaged by AFM in buffer. A 3D rendering representation is shown. (**h**) Force-distance curve obtained by indentation on the particle shown in (**g**). (**i**) The same virion imaged after indentation showing the point in the capsid where breakage occurred. The scale bars represent 100 nm. The AFM panels cover a 300 × 300 nm area

3.3 Negative Staining Immunoelectron Microscopy of Purified Viruses

1. Glow discharge collodion/carbon coated Ni grids.
2. Adsorb purified adenovirus particles to the carbon support by floating the grid on a 3–5 μL drop of the sample for 5 min (carbon side facing the grid) (*see* **Note 9**).
3. Blot by briefly touching the edge of the grid to a piece of Whatman paper. Do not allow the grid to become completely dry.
4. Wash by floating the grid on a drop of TBS for 2 min. Blot as above.
5. Wash and block by floating on a drop of TBG for 15 min (*see* **Note 10**). Blot.
6. Float on a drop of primary antibody at the required dilution (in TBS/BSA) for 50 min. Several different dilutions should be tested to optimize the ratio between specific labeling and background. In parallel, prepare grids labeled with the positive and negative control antibodies (*see* Fig. 2d, e). It is important to keep a high humidity environment by placing the experiment inside a petri dish containing wet paper all around the borders.
7. Blot and wash by floating on a drop of TBG, 10 min. Blot. Repeat twice.
8. Incubate on a drop of secondary antibody conjugated with 10 nm colloidal gold, diluted to 15 % v/v in TBS/BSA for 30 min. Blot.
9. Wash with TBG, 5 min. Blot. Repeat twice.
10. Wash with TBS, 5 min. Blot. Repeat twice.
11. Negative staining with 2 % uranyl acetate for 45 s. Blot and let dry (*see* **steps 9** and **10** in Subheading 3.2).

3.4 Cryo-Electron Microscopy

1. Glow-discharge Quantifoil grids.
2. Cool down the plunge freezing device using liquid nitrogen. Once the temperature in the plunging area is below −170 °C, fill the cryogen reservoir with ethane. Put some cryo-grid boxes inside the plunging area so they are precooled by the time the sample is vitrified (*see* **Note 11**).
3. Hold one Quantifoil grid with the tweezers. Load the tweezers on the plunging device and apply a 2–3 μL drop of adenovirus sample to the carbon side. Blot by applying a triangle of Whatman paper to the carbon face until most of the drop is removed and only a very thin layer of liquid remains. If using manual blotting, this step may require extensive hands-on practice until the desired layer thickness is achieved. Remove the paper and immediately release the plunging device, to avoid deleterious evaporation effects.

4. Transfer the grid from the ethane to a precooled cryo-box and store in liquid nitrogen until used. Make sure that once vitrified, the grid is never exposed to air or temperatures higher than −160 °C. Otherwise, undesirable effects due to temperature increase (*see* **Note 11**) or contamination by condensation may occur.
5. Cool down the electron microscope anticontaminator and the cryo-sample holder using a cryo-transfer device and liquid nitrogen. When the cryo-holder reaches a stable temperature below −170 °C, transfer the grid from the cryo-box to the sample holder using precooled tweezers.
6. Load the cryo-holder on the electron microscope. Wait for 30 min while monitoring that the holder temperature stays stable below −170 °C.
7. Observe the grid at low magnification (950×). Select grid squares with an adequate ice thickness and record their positions.
8. Set up the microscope to low dose imaging parameters, so that the sample areas of interest are irradiated exclusively during image recording while focusing is performed in an adjacent location.
9. Record micrographs at 60,000× magnification in low dose conditions, covering a defocus range between −1.5 and −3.5 μm (*see* Fig. 2f).
10. Digital format micrographs (either directly recorded if the microscope is equipped with a digital image acquisition system, or digitized in a high quality scanner such as the Nikon Coolscan 9000) are processed for viral particle extraction, alignment, and combination into a 3D reconstruction. The resolution attained in the final map will be related not only to the number of particles averaged but also to the quality of each individual image. A variety of specialized image processing software tools are available in the field [25]. Some popular ones for reconstruction of icosahedral virus particles are FREEALIGN [26], XMIPP [27], Bsoft [28], or IMIRIS [29] (*see* **Note 12**).

3.5 Atomic Force Microscopy

3.5.1 Sample Preparation

1. Cut a ~1 cm^2 piece from a mica sheet.
2. Cleave mica with tape to obtain a clean surface.
3. Dilute the adenovirus sample from a single-use aliquot in the solution of $NiCl_2$ in HBS to obtain a final solution of ~5×10^{11} vp/mL and 5 mM Ni^{2+}.
4. Place a 20 μL drop of final solution on the mica substrate.
5. Incubate for 30 min at 4 °C.
6. Wash the sample with a solution of 5 mM $NiCl_2$ in HBS.

3.5.2 AFM Setup and Imaging

1. Place the mica incubated with the virus on the AFM holder.
2. Immerse the sample in 500 μL of 5 mM $NiCl_2$ in HBS buffer (*see* **Note 13**).
3. Place the holder with the mica on a piezoelectric actuator, which controls the sample position.
4. Calibrate the cantilever spring constant (*see* **Note 14**).
5. Pre-wet the AFM tip with 30 μL of $NiCl_2$-HBS.
6. Focus the laser spot on the free extreme of the cantilever.
7. Tune the photodiode so that the laser spot is centered.
8. Proceed to bring the AFM tip closer to the surface of the sample.
9. Turn on Jumping Mode plus (JM+) provided by WSxM (*see* **Note 15**).
10. Optimize the measuring scanning parameters for best sample imaging (*see* **Note 16**). Scan large areas (about 3 × 3 μm) to find viral particles on the surface.
11. Once a viral particle has been located, scan an area of 300 × 300 nm and 128 × 128 pixels to have enough resolution to image a single viral particle (*see* Fig. 2g) (*see* **Note 17**).
12. If nanoindentation experiments are to be performed, obtain a force-distance curve on a substrate area next to the viral particle. This will provide the cantilever deflection characteristics in the absence of specimen.
13. Center on the viral particle and obtain a new force-distance curve indenting on the particle beyond its rupture point (*see* Fig. 2h).
14. Take a new image of the viral particle to observe the breakage pattern (*see* Fig. 2i).
15. Repeat for as many viral particles as needed to obtain adequate statistical coverage.

3.5.3 Image Analysis

The AFM image processing software WSxM stores several data files for the different features (topography, normal force, adhesion among others) recorded with JM+. The following points must be born in mind when analyzing the data:

1. Topography images usually present an artifactual tilt when the surface is not totally perpendicular to the z-scan. To eliminate this effect, computationally fit the surface with a plane from an area free from viral particles.
2. Changing the color scale and equalizing the data helps reveal details of the virion topography.

3. Lateral distances appear larger due to tip-sample dilation or to drift during acquisition.
4. Nanoindentation curves can be depicted in force vs. deformation plots by subtracting the curve of the cantilever deflection on substrate (*point 12* above) from the curve of the deformation of the cantilever-virus system. The nanoindentation curve provides information about the stiffness and breaking force of the viral capsid.

3.6 Electron Microscopy of Infected Cells

3.6.1 Fixation, Dehydration, and Embedding in Epoxy Resins for Morphological Studies

1. Prepare fixative (2 % glutaraldehyde, 1 % tannic acid in 0.4 M HEPES, pH 7.2). Dissolve 0.5 g of tannic acid in 46 mL of 400 mM HEPES, pH 7.2. Filter the solution with a 0.22 μm filter. Add 4 mL of EM-grade 25 % glutaraldehyde and protect from light. If the fixative is to be used the same day, keep it at room temperature. If it will be used the next day, store at 4 °C and filter before use.
2. Remove the medium from the culture plates, wash with PBS and add the fixative (enough quantity to cover the cells). However if many cells are detached, collect the medium and centrifuge for 5 min to pellet the cells. Wash with PBS and centrifuge again. Add the fixative and incubate 1–2 h at room temperature, never less than 1 h. Centrifuge and remove the fixative; then, add 1 mL of HEPES buffer. If the fixative was added to the culture plate you will need to use a cell scraper to detach and collect the cells. Centrifuge, remove the fixative, and add 1 mL of HEPES to the cell pellet. Store fixed cells at 4 °C until next step (*see* **Note 18**).
3. Transfer the pellet to a 1.5 mL Eppendorf tube. Wash the cells three times with 1.3 mL of HEPES for 15 min. Resuspend the cells using a wooden toothpick. Centrifuge at low speed for 5 min (minimal speed to precipitate cells, approx. 2,352 ×*g*).
4. *En bloc* staining: Add 300 μL of 1 % OsO_4 in 0.8 % $K_3[Fe(CN)_6]$ to the pellet and resuspend with a wooden toothpick. Incubate for 1 h at 4 °C, protected from the light. Wash three times with HEPES and centrifuge at low speed for 5 min. Add 300 μl of 2 % uranyl acetate and resuspend. Incubate for 40 min at 4 °C. Wash three times with HEPES and centrifuge. To make sure that all the staining agent is removed, incubate cells for 10 min with HEPES between washes.
5. Dehydration: prepare dilutions of dry acetone at 50 %, 70 %, and 90 % v/v in water. Add acetone 50 % to the cells, resuspend, and incubate for 15 min at 4 °C. Centrifuge at low speed for 5 min. Remove the supernatant, add acetone 70 %, resuspend, incubate for 15 min at 4 °C, and centrifuge (process can be paused for up to 1 h at this step if required). Repeat these steps once with acetone 90 % and twice with acetone 100 %.

6. Prepare 100 mL of Epon resin (*see* **item 9** in Subheading 2.6.1 and **Note 19**).
7. Embedding: prepare a mixture of acetone–Epon (1:1). Incorporate the mixture to cells and incubate overnight under gentle rotation. Centrifuge and remove the acetone–Epon mixture, then add only Epon (1 mL approx.). Incubate for 6 h at room temperature with agitation. When incubation time is over, centrifuge the samples, remove the resin, add fresh Epon, and incubate overnight with rotation (*see* **Note 20**).
8. Encapsulation: Centrifuge the samples and change Epon. Mix by rotation for 2 h. Cut small (0.7 × 0.4 cm) paper labels and write sample identification and date with a pencil. Put one label inside each beem capsule. Centrifuge the samples and remove approximately three quarters of the resin content. Resuspend the cells in the remaining resin and transfer them to the beem capsules. Pour enough cell–Epon mixture to fill the pyramidal part of each capsule. Depending on the amount of cells, 2 or 3 capsules can be filled per sample. Centrifuge to pellet down the cells (to centrifuge, put closed capsules into Eppendorf tubes) and completely fill the capsules with Epon until a convex meniscus is formed. Make sure that the labels are clearly visible.
9. Polymerization: incubate the samples at 60 °C for 48 h without closing the beem capsules. Finally, the beem capsule is cut open with a razor and removed to obtain a block ready for ultramicrotomy (*see* **Note 21**).

3.6.2 Ultramicrotomy

1. Load a resin block in the ultramicrotome holder and place in the trimming block with the cell pellet facing upwards.
2. Trim away resin from the surface of the block tip using a single edge razor blade, until you reach the cell pellet.
3. Prepare glass knives by cutting a glass stripe according to the knife maker instruction manual.
4. Remove the sample holder from the trimming block and load it in the microtome arm. Place a glass knife in the microtome knife holder.
5. Trim the block surface by cutting 1 μm thick slices with the glass knife. Using forceps or a pair of needles, take some sections and put in a drop of water on a microscope slide. Heat to evaporate the water. Cover the section with a drop of toluidine blue, heat on a warm plate for 30–120 s. Gently rinse with distilled water to eliminate excess stain. Dry and observe in an optical microscope. Select a region with many cells and locate this area in the Epon block.

6. Place the block holder back into the ultramicrotome trimming block.
7. With the blade, trim the block tip to shape it as a trapeze containing the cell-containing selected region (0.5 × 1.0 mm approx.).
8. Remove the sample holder from the trimming block and load it in the microtome arm. Place a new glass knife in the microtome knife holder.
9. Trim the block surface by cutting 40 nm thick slices to polish until the surface shines.
10. Cut 70 nm thick sections using a diamond knife or a glass knife with an attached plastic boat filled with distilled water. Fill the boat until the water surface shines like a mirror. This is important because in these conditions the thin sections floating on the water will reflect a particular color that indicates their thickness. With the loop, pick a few silver or golden sections and transfer them to a grid. If necessary, sections can be moved around on the water surface to regroup them using a hair glued to a wooden stick.

3.6.3 Section Staining

1. Centrifuge saturated uranyl acetate and lead citrate for 5 min at 18,550 × *g*.
2. Put a drop (25 μL) of uranyl acetate on a piece of Parafilm and place the grid on the drop with the sections in direct contact with the solution. Incubate for 25 min in a dish covered to protect the uranyl acetate from light.
3. Blot the grid with Whatman filter paper to remove the uranyl acetate (but do not allow it to dry completely).
4. Wash the grid quickly by consecutive immersion in four large drops (300 μL each) of MilliQ water.
5. Dip the grid in a drop of lead citrate (200 to 100 μL) and incubate for 90 s. Make sure that the grid is completely immersed in the drop and cover the plate during the incubation to minimize precipitation of lead citrate due to contact with oxygen.
6. Wash the grid quickly in four drops (300 μL) of MilliQ Water.
7. Blot the grid and place on a new piece of filter paper inside a petri dish, with the sections facing up. Wait until it is dry before imaging at the electron microscope (*see* Fig. 3a).

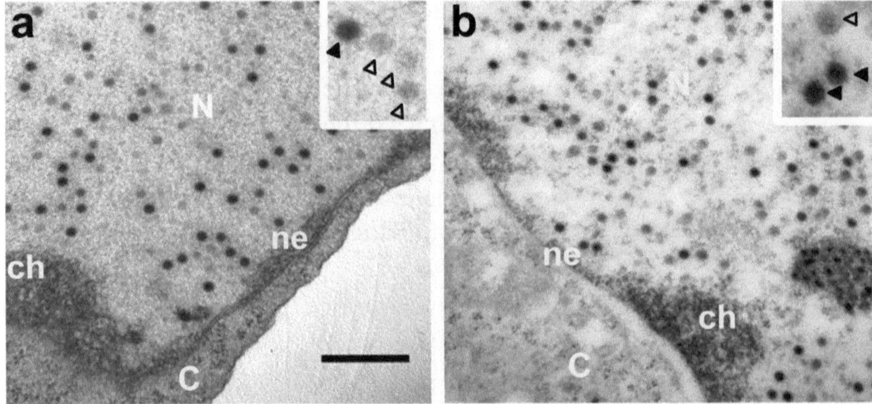

Fig. 3 Ultrathin sections of adenovirus-infected cells. A549 cells were infected with Ad2 ts1 mutant and processed for embedding after 36 hpi. (**a**) Room temperature dehydration and Epon embedding. (**b**) Freeze-substitution. The insets show details of full (*filled arrowheads*) and empty (*hollow arrowheads*) capsids imaged within the cell nucleus. Notice the different contrast obtained by the two methods, particularly for membranes (*black* in Epon, *white* in freeze-substitution). *C* cytoplasm, *ch* chromatin, *N* nucleus, *ne* nuclear envelope. Scale bar, 0.5 μm

4 Notes

1. Nickel grids must be used for "on grid" immunogold labeling, instead of copper ones, due to the reactivity of the copper in saline solutions during prolonged incubations.
2. In the market there are gold particles of different sizes, allowing the localization of two different epitopes if necessary. Another alternative is the use of gold conjugated protein A instead of a secondary antibody (for example Protein A EM-grade 10 nm Cat#25284 Electron Microscopy Sciences).
3. The presence of glycerol or sucrose in buffers results in diminished contrast when imaging unstained specimens in frozen-hydrated conditions, due to the similarity of electron density between these cryo-protectants and the biological material.
4. Although other electron sources can also be used for cryo-EM imaging, a Field Emission Gun (FEG) equipped microscope provides the best quality data to reach subnanometer resolution, due to the improved coherence of the electron beam generated.
5. The choice of an appropriate substrate is crucial in AFM studies. The interaction between surface and sample plays a big role in aspects such as attachment or structural preservation. All the experiments must be carried out with the same surface since particles may behave differently due to the interactions with the surface. For example, adenovirus is unstable on HOPG (Highly Ordered Pyrolytic Graphite), but it is well attached on

mica (provided that bivalent cations are present in the buffer solution) or functionalized glass.

6. The total volume of sample depends on the particular spectrofluorimeter and cuvette models used. The requirement is that the sample lies in the optical path.
7. Double distilled water can also be used for washing, but it may be damaging to the structures due to osmotic shock, and may induce staining artifacts (e.g., positive staining).
8. A common problem in negative staining is its variability: even within the same grid square, negative and positive staining areas may be observed. Positive staining arises when uranyl acetate binds to the biological material, instead of being excluded. In positively stained areas, viral particles appear dark on a light background, and structural details are no longer visible (*see* Fig. 2c). These areas must not be used for structural studies. Uranyl acetate usually works well for adenovirus, but if needed, alternative staining agents such as ammonium molybdate can be used. Ammonium molybdate is usually prepared as a 2 % w/v solution in double-distilled water (pH ~ 5.3) and it can be titrated with NaOH or NH_4OH to pH 6.0–7.0.
9. The virus sample should appear homogeneously spread all over the grid, allowing observation of isolated particles. An optimal virus concentration would be 1×10^{11} vp/mL.
10. TBS/BSA and TBG can be stored in aliquots at −20 °C. TBG has to be thawed at 37 °C to melt the gelatin.
11. It is crucial that any tool, storage element, etc. that will be in contact with the grid after vitrification is precooled in liquid nitrogen. Otherwise the sample temperature will rise and undesirable ice crystal formation, or even thawing, will occur.
12. As an alternative to averaging, cryo-electron tomography allows recovery of the 3D structure of individual viral particles without averaging, providing information on unique structural features [30]. For cryo-electron tomography, the viral sample is mixed with 10-nm colloidal gold particles (AURION, Wageningen, The Netherlands) before plunge freezing. The electron microscope must be equipped with a high-tilt cryo holder and an automated image acquisition system [31], to take a tilt series of the same field covering an angular range between −70 and +70 °C. The images in the tilt series are aligned using the gold particles as fiducials, normalized and reconstructed using specialized software packages [32–34].
13. A large volume of buffer reduces the risk of high variations of ionic strength due to evaporation, ensuring virus stability throughout the experiment.
14. Among the available methods [35], the one proposed by Sader [36] works very well for the cantilevers used here.

15. Among all the modes of measurement provided by WSxM, Jumping Mode plus (JM+) is optimal to image adenovirus because very low, accurate normal forces are applied to avoid damage of the sample by the scanning tip [37].
16. Adjust scanning parameters: (a) Set the *setpoint* to measure about 0.05 V. This will determine the force that the microscope applies on the sample. The applied force depends on the sensitivity of the microscope, that it is in turn related to the laser spot position on the cantilever and the photodiode. (b) Set *Jump off* to a height slightly larger than the half height of the viral particle (i.e., 45 nm for adenovirus). (c) Set *Jump sample* to approx. twice the value of the *jump off*. (d) Set *control cycles* to a number high enough to let the feedback act (i.e., 20). (d) To use JM+, turn on the *Stop moving if limit is reached* box and use the parameters: *limit(V)* higher than the measurement set point. The *dragging factor* suggested by WSxM usually works well. Feedback parameters depend on the piezo actuator and cantilever.
17. Although resolution depends on the tip shape, it should be possible to resolve the hexon trimers with the conditions described here.
18. If needed, cells can be prepared for sectioning in their monolayer disposition rather than in a pellet. Autoclave circular glass coverslides and incubate on a drop of poly-lysine (use a 100 μL drop for a round cover slide of 12 mm diameter) for 30 min at 37 °C. Wash the coverslide with PBS and place in a culture plate (12-well plates are recommended). The poly-lysine keeps the cells attached to the glass throughout the subsequent procedures. Follow the steps described in Subheading 3.6.1 for fixation and embedding, but use ethanol instead of acetone for dehydration. For encapsulation, use gelatin instead of beem capsules. Fill the longest half of the capsule with Epon and place it upside down on the glass slide, which should be covered by a thin layer of Epon. Polymerize for 48 h at 60 °C. For ultramicrotomy, the glass is removed using a rapid freeze-thawing procedure, transferring the capsule and attached glass quickly from liquid nitrogen to hot water several times.
19. Epon can be prepared in advance, with the exception of BDMA that must be incorporated on the same day of use. The mixture of resins (without BDMA) can be kept under gentle rotation over night to mix well.
20. A new batch of fresh Epon without BDMA can be prepared at this point and left under rotation overnight for use the following day.
21. An alternative to the standard Epon embedding methodology is *freeze substitution*. Freeze substituted samples are dehydrated

and embedded at low temperatures to minimize extraction and better preserve the native structure of the samples, including immunogenic characteristics for labeling (*see* Fig. 3b). Specific materials needed for freeze substitution are:

(a) Fixative: Paraformaldehyde 4 % in PBS (protect from light).

(b) Leica EM CPC cryofixation system.

(c) Leica EM AFS2 freeze substitution and low temperature embedding system.

(d) Embedding medium: Lowicryl HM20 embedding kit (Agar Scientific cat#R1034).

(e) En bloc staining and dehydration agent: 0.5 % Uranyl acetate in dried methanol (protect from light and store at −20 °C).

The general procedure is similar to that of Epon embedding, with the appropriate reactive changes to carry out a milder chemical fixation and dehydration. After chemical fixation, samples are treated with a cryo-protectant like glycerol to prevent the formation of ice crystals in the samples during freezing. Cell pellets are flash-frozen in the Leica EM CPC by plunging in liquid ethane. Then, the samples are transferred to the freeze-substitution unit (Leica AFS) for low temperature dehydration, embedding, and polymerization with UV light. The protocol is summarized in Table 1, and the details of AFS programming are given in Table 2.

Table 1
Summary of freeze-substitution protocol

Step	Reagent	Time	Temperature
Fixation	Paraformaldehyde 4 % in PBS	10 min	Room temperature
Cryo-protection	Glycerol 15 % in PBS Glycerol 30 % in PBS	15 min 15 min	4 °C 4 °C
Cryo-fixation by plunging			
Staining	Uranyl acetate 0.5 % in methanol	Change twice daily during 3 days	−90 °C
Wash	Methanol 100 %	1 h, repeat three times	−40 °C
Embedding	Methanol: Resin (3:1) Methanol: Resin (1:1) Methanol: Resin (1:3) Resin 100 % Resin 100 % Resin 100 % Resin 100 %	1 h 1 h 1 h 15 min Overnight 4 h 48 h	−40 °C

Table 2
Leica AFS programming for freeze substitution

Step	Temp start (°C)	Temp end (°C)	Time	Reagent	UV
1	-90	-90	60:00:00	Methanol/uranyl acetate	No
2	-90	-40	07:00:00	Methanol/uranyl acetate	No
3	-40	-40	31:00:00	Methanol Resin dilutions Resin 100 %	No
4	-40	-40	48:00:00	Resin 100 %	Yes
5	-40	20	04:00:00	Resin 100 %	Yes
6	20	20	48:00:00	Resin 100 %	Yes

The amounts to prepare 20 g of HM20 resin are:

(a) 2.98 g Cross-linker D

(b) 17.02 g Monomer E

(c) 0.1 g Initiator C

Unlike Epon, the complete HM20 mixture can be prepared and stored at −20 °C until use. Also the uranyl acetate in methanol can be prepared and stored at −20 °C. All solutions or materials have to be cooled down in the AFS for 15 min prior to them entering in contact with the samples.

Acknowledgements

Work supported by grants from the Ministry of Science and Innovation of Spain BFU2010-16382/BMC to C. S. M.; MAT2008-02533, PIB2010US-00233, and FIS2011-29493 to P. J. P; FIS2010-10552-E and FIS2011-16090-E to C. S. M. and P. J. P.; as well as from the Local Madrid Government P2009/MAT-1467 to P. J. P. A. J. P. B. holds a Juan de la Cierva postdoctoral contract (JCI-2009-05187) from the Ministry of Science and Innovation of Spain. A. O. E., G. N. C. and R. M. C. are recipients of predoctoral fellowships from the Ministry of Education, the CSIC-JAE program (JAEPre_09_00041), and the Instituto de Salud Carlos III of Spain (FI08/00035), respectively. The technical assistance of María López and María Angeles Fernández is acknowledged, as well as constant support by members of the EM Service and the Department of Macromolecular Structures at the CNB-CSIC. Protocols described in this chapter are by no means the contribution of a single research group. They have been developed and refined over time by contributions from the scientific community at large.

References

1. Rexroad J, Martin TT, McNeilly D, Godwin S, Middaugh CR (2006) Thermal stability of adenovirus type 2 as a function of pH. J Pharm Sci 95:1469–1479
2. Pérez-Berná AJ, Ortega-Esteban A, Menéndez-Conejero R, Winkler DC, Menéndez M, Steven AC et al (2012) The role of capsid maturation on adenovirus priming for sequential uncoating. J Biol Chem 287:31582–31595
3. Harris JR (2007) Negative staining of thinly spread biological samples. In: Kuo J (ed) Electron microscopy. Methods in molecular biology™. Humana Press, Totowa, New Jersey, pp 107–142
4. Horne RW, Brenner S, Waterson AP, Wildy P (1959) The icosahedral form of an adenovirus. J Mol Biol 1:84–86
5. Hayat MA (2000) Principles and techniques of electron microscopy. biological applications, 4th edn. Cambridge University Press, Cambridge
6. Adrian M, Dubochet J, Lepault J, McDowall AW (1984) Cryo-electron microscopy of viruses. Nature 308:32–36
7. Dobro MJ, Melanson LA, Jensen GJ, McDowall AW (2010) Chapter three—plunge freezing for electron cryomicroscopy. In: Jensen Grant J (ed) Methods in enzymology. Academic Press, New York, pp 63–82
8. Sun J, Li H (2010) Chapter ten—how to operate a cryo-electron microscope. In: Jensen Grant J (ed) Methods in enzymology. Academic Press, New York, pp 231–249
9. Baker TS, Olson NH, Fuller SD (1999) Adding the third dimension to virus life cycles: three-dimensional reconstruction of icosahedral viruses from cryo-electron micrographs. Microbiol Mol Biol Rev 63:862–922
10. Grigorieff N, Harrison SC (2011) Near-atomic resolution reconstructions of icosahedral viruses from electron cryo-microscopy. Curr Opin Struct Biol 21:265–273
11. Stewart PL, Burnett RM, Cyrklaff M, Fuller SD (1991) Image-reconstruction reveals the complex molecular-organization of adenovirus. Cell 67:145–154
12. Liu H, Jin L, Koh SB, Atanasov I, Schein S, Wu L et al (2010) Atomic structure of human adenovirus by cryo-EM reveals interactions among protein networks. Science 329:1038–1043
13. Reddy VS, Natchiar SK, Stewart PL, Nemerow GR (2010) Crystal structure of human adenovirus at 3.5 A resolution. Science 329: 1071–1075
14. San Martín C (2012) Latest insights on adenovirus structure and assembly. Viruses 4: 847–877
15. Binnig G, Quate CF, Gerber C (1986) Atomic force microscope. Phys Rev Lett 56:930–933
16. Giessibl FJ (1995) Atomic-resolution of the silicon (111)-(7x7) surface by atomic-force microscopy. Science 267:68–71
17. Alessandrini A, Facci P (2005) AFM: a versatile tool in biophysics. Meas Sci Technol 16: R65–R92
18. Ivanovska IL, De Pablo PJ, Ibarra B, Sgalari G, MacKintosh FC, Carrascosa JL et al (2004) Bacteriophage capsids: tough nanoshells with complex elastic properties. Proc Natl Acad Sci USA 101:7600–7605
19. Schaap IAT, Carrasco C, De Pablo PJ, MacKintosh FC, Schmidt CF (2006) Elastic response, buckling, and instability of microtubules under radial indentation. Biophys J 91:1521–1531
20. Sotres J, Baró AM (2010) AFM imaging and analysis of electrostatic double layer forces on single DNA molecules. Biophys J 98: 1995–2004
21. Carrasco C, Luque A, Hernando-Pérez M, Miranda R, Carrascosa JL, Serena PA et al (2011) Built-in mechanical stress in viral shells. Biophys J 100:1100–1108
22. Hernando-Pérez M, Miranda R, Aznar M, Carrascosa JL, Schaap IAT, Reguera D et al (2012) Direct measurement of phage phi29 stiffness provides evidence of internal pressure. Small 8:2366–2370
23. Roos WH, Bruinsma R, Wuite GJL (2010) Physical virology. Nat Phys 6:733–743
24. Horcas I, Fernández R, Gómez-Rodriguez, JM, Colchero J, Gómez-Herrero J, Baró AM (2007) WSXM: A software for scanning probe microscopy and a tool for nanotechnology. Rev Sci Instrum 78:013705
25. Voss NR, Potter CS, Smith R, Carragher B (2010) Chapter fifteen—software tools for molecular microscopy: an open-text wikibook. In: Jensen Grant J (ed) Methods in enzymology. Academic Press, New York, pp 381–392
26. Grigorieff N (2007) FREALIGN: high-resolution refinement of single particle structures. J Struct Biol 157:117–125
27. Scheres SH, Núñez-Ramírez R, Sorzano CO, Carazo JM, Marabini R (2008) Image processing for electron microscopy single-particle analysis using XMIPP. Nat Protoc 3:977–990
28. Heymann JB, Belnap DM (2007) Bsoft: image processing and molecular modeling for electron microscopy. J Struct Biol 157:3–18
29. Liang Y, Ke EY, Zhou ZH (2002) IMIRS: a high-resolution 3D reconstruction package integrated with a relational image database. J Struct Biol 137:292–304
30. Chang J, Liu X, Rochat RH, Baker ML, Chiu W (2012) Reconstructing virus structures from nanometer to near-atomic resolutions with

cryo-electron microscopy and tomography. Adv Exp Med Biol 726:49–90

31. Zheng SQ, Sedat JW, Agard DA (2010) Chapter twelve—automated data collection for electron microscopic tomography. In: Jensen Grant J (ed) Methods in enzymology. Academic Press, New York, pp 283–315
32. Heymann JB, Cardone G, Winkler DC, Steven AC (2008) Computational resources for cryo-electron tomography in Bsoft. J Struct Biol 161:232–242
33. Scheres SH, Melero R, Valle M, Carazo JM (2009) Averaging of electron subtomograms and random conical tilt reconstructions through likelihood optimization. Structure 17:1563–1572
34. Kremer JR, Mastronarde DN, McIntosh JR (1996) Computer visualization of three-dimensional image data using IMOD. J Struct Biol 116:71–76
35. Burnham NA, Chen X, Hodges CS, Matei GA, Thoreson EJ, Roberts CJ et al (2003) Comparison of calibration methods for atomic-force microscopy cantilevers. Nanotechnology 14:1–6
36. Sader JE, Chon JWM, Mulvaney P (1999) Calibration of rectangular atomic force microscope cantilevers. Rev Sci Instrum 70: 3967–3969
37. Ortega-Esteban A, Horcas I, Hernando-Pérez M, Ares P, Pérez-Berná AJ, San Martín C et al (2012) Minimizing tip–sample forces in jumping mode atomic force microscopy in liquid. Ultramicroscopy 114:56–61
38. San Martín C, Glasgow JN, Borovjagin A, Beatty MS, Kashentseva EA, Curiel DT et al (2008) Localization of the N-terminus of minor coat protein IIIa in the adenovirus capsid. J Mol Biol 383:923–934

Chapter 2

Proteome Analysis of Adenovirus Using Mass Spectrometry

Sara Bergström Lind, Konstantin A. Artemenko, and Ulf Pettersson

Abstract

Analysis of proteins and their posttranslational modifications is important for understanding different biological events. For analysis of viral proteomes, an optimal protocol includes production of a highly purified virus that can be investigated with a high-resolving analytical method. In this *Methods in Molecular Biology* paper we describe a working strategy for how structural proteins in the Adenovirus particle can be studied using liquid chromatography–high-resolving mass spectrometry. This method provides information on the chemical composition of the virus particle. Further, knowledge about amino acids carrying modifications that could be essential for any part of the virus life cycle is collected. We describe in detail alternatives available for preparation of virus for proteome analysis as well as choice of mass spectrometric instrumentation suitable for this kind of analysis.

Key words Adenovirus, Proteome analysis, Posttranslational modification, Mass spectrometry, LC-MS

1 Introduction

Human adenovirus consists of about a dozen of proteins that assemble into a dense structure [1, 2]. Even though the amino acid sequence of the proteins and their copy numbers per virus particle are well known for many adenovirus species [3, 4] as well as the atomic structure of adenovirus and some adenovirus capsid proteins [5–9], there is still much to reveal. Proteome analysis of purified virus particles using highly sensitive mass spectrometric analysis adds a layer of posttranslational modifications (PTMs) to the structural information. It is also possible to identify new proteins, encoded by the virus genome, attached or integrated in the virion. Today, liquid chromatography–mass spectrometry (LC-MS) is the method of choice for large-scale characterization of proteins, peptides, and PTMs. The so-called shotgun proteomics method provides unbiased protein identification and site determination of PTMs [10–12]. Proteome analysis of viruses requires highly purified virions, which in the case of non-enveloped viruses can be

Miguel Chillón and Assumpció Bosch (eds.), *Adenovirus: Methods and Protocols*, Methods in Molecular Biology, vol. 1089, DOI 10.1007/978-1-62703-679-5_2, © Springer Science+Business Media, LLC 2014

produced by CsCl gradient centrifugation. In this way, the virus is separated from host cell proteins. Further, viral structural proteins are extracted from the virus and cleaved into peptides. For example, in phosphoproteome analysis, enrichment of phosphopeptides before mass spectrometric analysis improves the results by reducing the competition from non-phosphorylated species [13] and is a requirement for detecting phosphorylation sites in complex samples as cell lysates [14]. If studying a fractionated part of the cellular proteome PTMs, e.g., acetylation and methylation, and to some extent also phosphorylations, it can be analyzed without preceding enrichment [15]. The same principle applies to analysis of the relatively small viral proteome. Analysis of peptides carrying a PTM with MS requires detection of a mass shift corresponding to the added modification, e.g., +79.98 Da for phosphorylation, and localization of the PTM site in the peptide chain. Technically, this requires high mass accuracy and resolution to detect the mass shift and tandem MS analysis (MS/MS) of the precursor ion to localize the site of modification. With the aid of search engines and databases the recorded MS/MS spectra are converted to peptide sequences, with possible PTMs, that also are matched to protein identity. We have used mass spectrometry and shown that there are at least 29 different sites of phosphorylation in 11 different proteins in the adenovirus type 2 (Ad2) particle [16]. In agreement with previous studies [17–21] we showed that the pIIIa protein is the most extensively phosphorylated protein and for this protein 12 sites of phosphorylation were identified [16]. Below we present the key steps for proteome analysis of adenovirus by LC-MS summarized in a protocol.

2 Materials

For quality of chemicals, *see* **Note 1**.

2.1 Adenovirus Production

1. HeLa spinner cells (ATCC/LCG).
2. Medium for cell growth: Minimal Essential Medium (MEM), spinner modification supplemented with 10 % fetal bovine serum, 1 % Penicillin-Streptomycin (PEST), 1 % L-glutamine. Store prepared medium at 4 °C (up to 2 months) and warm to 37 °C in water bath upon use. To obtain a virus preparation, around 5 L of medium is needed.
3. Medium for infection: MEM, spinner modification supplemented with 1 % PEST and 1 % L-glutamine. Store prepared medium at 4 °C (up to 2 months) and warm to 37 °C in water bath upon use.
4. Adenovirus (ATCC/LCG).

2.2 Virus Purification

1. Storage buffer for virus: 10 mM Tris–HCl, pH 7.9. Prepare this solution fresh as required by diluting from a stock solution of 1 M.
2. Purification reagents: 5 % (w/v) sodium deoxycholate, make fresh as required.
3. CsCl solution of density 1.41 g/cm^3: Weight 27.42 g CsCl and fill to 50 mL with 10 mM Tris–HCl pH 7.9. Sterile filtrate the CsCl solution by 0.22 μm filters. Densities can be double-checked by weighing. All CsCl solutions are long term stable in room temperature.
4. CsCl solution of density 1.34 g/cm^3: Weight 22.71 g CsCl and fill to 50 mL with 10 mM Tris–HCl pH 7.9. Sterile filtrate the CsCl solution by 0.22 μm filters.
5. CsCl solution of density 1.27 g/cm^3: Weight 18.47 g CsCl and fill to 50 mL with 10 mM Tris–HCl pH 7.9. Sterile filtrate the CsCl solution by 0.22 μm filters.
6. For dialysis: Slide-A-Lyzer Cassette (10 kDa MWCO, 0.5–3 or 3–12 mL, Pierce/Thermo Scientific Inc.) and 10 mM Tris–HCl, pH 8.0.
7. Ultra-clear ultracentrifugation tubes (344058, Beckman Coulter).
8. AH-429 rotor and ultracentrifuge.

2.3 Preparation of Samples for Analysis of the Viral Proteome by Mass Spectrometry

2.3.1 Common

1. Phosphatase inhibitors can be prepared from the following stock solutions:
   (a) 500 mM NaF (500×), prepare in water, aliquot and store in freezer.
   (b) 1 M β-glycerophosphate (500×), prepare in water, aliquot and store at −20 °C. Thawed aliquots can be stored at 4 °C for 6 months.
   (c) 500 mM Na-pyrophosphate (500×), prepare in water, store at 4 °C for up to 1 month.
   (d) 500 mM Na_3VO_4 (500×) needs to be activated/depolymerized. Add powder and fill to 90 % of desired volume with water. Adjust pH to 10 using 1 M HCl. At this pH the solution becomes yellow. Boil the solution until it is colorless and put on ice for cooling. Readjust the pH to 10 and repeat boiling and cooling until the solution remains stably colorless at pH 10. Adjust to desired volume with water, aliquot and store at −20 °C.
2. 1–1.25 M dithiothreitol (DTT) Aliquot and store at −20 °C.
3. Iodacetoamide (IAA): Purchase as powder, weigh desired amount each time a solution should be prepared in Milli-Q water.
4. Acetonitrile.
5. EZQ protein quantitation kit (Molecular probes) or similar.

2.3.2 In-Solution Digestion

1. 200 mM HEPES buffer (10×).
2. Lysis buffer: 9 M urea, 20 mM HEPES, 2.5 mM sodium pyrophosphate, 1 mM β-glycerophosphate, and 1 mM sodium vanadate. Weigh desired amount of urea and dissolve in a small volume of water. Heat in water bath (60 °C for 5 min) and then add a virus aliquot to this solution. Further, add small volumes of stock solutions directly to the sample and dilute with Milli-Q water to a final volume for the desired concentrations.
3. Trifluoroacetic acid (TFA).
4. Trypsin (or other protease, e.g., LysC or chymotrypsin of sequencing grade). Enzymes are provided as powder. Dissolve in 1 mM HCl at a concentration of 1 μg/μL. Aliquot 1 μg/vial and store at −20 °C.
5. 10 kDa MWCO ultrafiltration unit (Millipore or similar).
6. Empore Disks C_{18} membrane (Varian).

2.3.3 (1D) SDS-PAGE Separation

1. Sample loading buffer for 1D SDS-PAGE (2× Laemmli sample buffer): 62.5 mM Tris–HCl pH 6.8, 2 % sodium dodecyl sulfate (SDS), 25 % glycerol, 0.1 % bromophenol blue or (5× Laemmli sample buffer): 156 mM Tris–HCl pH 6.8, 5 % SDS, 62.5 % glycerol, 0.025 % bromophenol blue. Add 1/20 or 1/8 of β-mercaptoethanol (v/v) to the 2× or 5× Laemmli sample buffer, respectively, as required.
2. Molecular weight marker: There are different markers available provided as a powder. Store powder at 4 °C until use and then add 2× Laemmli sample buffer with 1/20 β-mercaptoethanol (add the volume recommended by supplier) and heat to 95 °C for 5 min. Allow to cool on ice and then aliquot and store at −20 °C.
3. Running buffer: (a) If using tris-glycine gels (purchased or cast in-house), prepare SDS-running buffer consisting of 25 mM Tris base, 192 mM glycine, 0.1 % (w/v) SDS. Store at room temperature. (b) If using other SDS-PAGE systems, e.g., NuPAGE Bis-Tris Precast gels (Life Technologies) 4–12 %, then use the matching running buffer, e.g., NuPAGE MOPS SDS running buffer. Store at room temperature.
4. Fixation solution: 10 % methanol, 7 % acetic acid, 83 % water, store at room temperature.
5. Colloidial Coomassie staining, stock solution: 0.1 % CBB (Coomassie brilliant blue)-G250, 1 % phosphoric acid, 10 % w/v ammonium sulfate in water. Store at 4 °C, add 20 % methanol upon use.

2.3.4 (2D) SDS-PAGE Separation

1. Sample preparation solution for 2D electrophoresis: 7 M Urea, 2 M thiourea, 4 % chaps, 30 mM Tris buffer pH 8.5, 1 mM

Na_3VO_4, 1 mM NaF, 2 mM β-glycerophosphate, 1 mM Na-pyrophosphate and one tablet of protease inhibitor Complete Mini-EDTA-free (Roche Diagnostics) per 10 mL buffer.

2. Immobilized pH gradient (IPG) buffer: IPG buffer pH 4–7 and IPG buffer pH 3–11 NL (GE Healthcare).
3. IPG strips pH 4–7 and pH 3–11.
4. Ultra-clear ultracentrifugation tube (344062, Beckman coulter).
5. SW60Ti rotor and ultracentrifuge.
6. Rehydration solution: 8 M urea, 2 % (w/v) chaps, 0.5 % (v/v), IPG buffer, 0.002 % bromophenol blue.
7. Agarose sealing solution: 0.5 % agarose NA (GE Healthcare), 0.002 % bromophenol blue stock solution in SDS running buffer (*see* Subheading 2.3.3). Add all reagents and carefully heat the solution in a microwave until the agarose is completely dissolved. Dispense 1.5 mL aliquots to be stored at 4 °C. The solution is long-term stable. Upon use, boil and keep at 60 °C before sealing.
8. SDS equilibration buffer solution: 6 M urea, 75 mM Tris–HCl pH 8.8, 29.3 % (v/v) glycerol, 2 % (w/v) SDS, 0.002 % bromophenol blue. This is just a stock solution, upon use prepare (a) SDS equilibration buffer supplemented with 0.04 mM DTT and (b) SDS equilibration buffer supplemented with 0.25 g IAA per 10 mL buffer.
9. Mineral oil/Dry Strip cover fluid.
10. Solutions 3–5 specified above for *1D SDS-PAGE separation*.
11. Immobiline rehydration tray.
12. Ettan IGPhor isoelectrical focusing unit (or similar equipment).

2.3.5 In-Gel Digestion (Suitable for Excised Pieces from Both 1D and 2D SDS-PAGE)

1. Trypsin (or other protease, e.g., LysC or chymotrypsin). Enzymes are provided as powders. Dissolve in 1 mM HCl at a concentration of 1 μg/μL. Aliquot 1 μg/vial and store at −20 °C.
2. 50 mM NH_4HCO_3: Prepare fresh from stock solution of 1 M.
3. Acetonitrile.

2.3.6 Phosphopeptide Titanium Dioxide Enrichment

1. Loading buffer A: 0.4 % TFA, 80 % acetonitrile, and 19.6 % Milli-Q water.
2. Loading buffer B: 25 % lactic acid, 60 % acetonitrile, 0.3 % TFA, 14.7 % Milli-Q water.
3. Elution buffer A: 5 % ammonia solution and 95 % Milli-Q water.

4. Elution buffer B: 5 % pyrrolidine and 95 % Milli-Q water.
5. TiO_2 material (GL Sciences Inc., Tokyo, Japan).

2.4 LC-MS Analysis

1. Mobile phase buffer A: 0.5 % acetic acid in Milli-Q water.
2. Mobile phase buffer B: 0.5 % acetic acid, 89.5 % acetonitrile, and 10 % Milli-Q water.
3. Sample loading buffer: 0.5 % TFA in Milli-Q water.
4. Mass spectrometer, e.g., 7-tesla LTQ-FT Ultra tandem mass spectrometer (Thermo Fischer Scientific) equipped with a nano electrospray ion source (Proxeon Biosystems).
5. LC system, e.g., Agilent 1100 Nanoflow system.
6. Packing material for LC column: Reprosil-Pur C18-AQ, 3 μm resin (Dr. Maisch GmbH).
7. Fused silica emitter.

3 Methods

In order to perform proteome analysis of pure virions it is important to separate out host cell proteins that otherwise would interfere and in the worst case would dominate in the detection. CsCl gradient ultracentrifugation is an efficient method for adenovirus purification as described by Pettersson and Sambrook [22]. Virus is banded in a step gradient followed by equilibrium centrifugation. Purity of virus samples can be checked in a SDS-PAGE and Coomassie staining. Before MS detection the proteins in the virions need to be extracted and digested into peptides. Ideally, viral proteins should pass a minimal number of preparation steps before MS detection, since each step causes losses of material. On the one hand, in-solution digestion followed by a desalting step is the most straight-forward procedure. On the other hand LC-MS analysis is based on separation of digested peptides on a reversed-phase column and recording of mass spectra of eluted peptides. Eluting peptides are competing with each other in order to be properly detected by mass spectrometry. Therefore preliminary fractionation is often needed and then it is favorable to initially separate the proteins in a sample by 1D or 2D SDS-PAGE and then analyze selected protein bands or spots with MS. Aspects on expected detectability of Adenovirus structural proteins are discussed in **Note 2** and Table 1.

Similarly, for PTM analysis, enrichment is highly valuable in order to decrease the competition between modified and non-modified peptides. Most of posttranslationally modified peptides (or original proteins) can be enriched by immunoaffinity methods using PTM specific antibodies attached to magnetic beads [23], resin beads or columns [24]. This approach has been used for

Table 1
Total protein and phosphorylation analysis of proteins in the Adenovirus type 2 virion using mass spectrometry

Abbreviation	Protein name	Number per virion[a]	MS detection[b]	Sequence coverage (%)[c]	Number of phosphorylation sites detected[d]
pII	Hexon	720	22	37	3
pIII	Penton base	60	3	17	1
pIIIa	Capsid protein precursor pIIIa	60	14	41	12
pIV	Fiber	36	7	24	1[e]
pV	Core protein V/major core protein	157 ± 1	8	23	2
52K	Encapsidation protein 52K		2	12	2
pVI	Capsid protein precursor pVI	360	5	36	2
pVII	pVII protein	833 ± 19	2	14	2
pVIII	Capsid protein precursor pVIII	127 ± 3	3	22	3
pIX	Capsid protein IX	240	1	9	1
pTP	DNA terminal protein	2	3	6	1
pX/pμ	Late L2 mu core protein	125–160	–	–	–
Protease	Protease	~20	5	19	–

[a]Combined data from Flint et al. [1] and Stewart et al. [3]
[b]Number of unique tryptic peptides detected in total protein analysis of the Ad2 particle with $p < 0.05$
[c]Sequence coverage of proteins after total protein analysis of the Ad2 particle
[d]Verified phosphorylation sites, additional phosphorylated sequences might have also been observed [16]
[e]Only phosphorylated sequence detected, phosphosite not revealed

many PTMs, namely, acetylated [25] and ubiquitinated lysines [26], glycosylation (using lectin affinity columns) [27], and tyrosine phosphorylation [23]. Overall, phosphorylation is one of the most commonly studied PTMs, and enrichment of phosphorylation can be obtained in many ways. Apart from antibody-based affinity techniques, chemical affinity is widely used, e.g., immobilized metal affinity chromatography (IMAC), metal oxide affinity chromatography (MOAC), and titanium dioxide as reviewed by Dunn et al. [28]. Moreover, chemical modifications with further surface enrichment have also been developed based on phosphoramidate chemistry (PAC) [29] and Michael addition [30]. Below we describe the titanium dioxide affinity approach from GL Sciences Inc., which we have used for identifying phosphorylation sites in the proteome of Ad2 virions. As reviewed by Thingholm et al. there are other alternatives than those described here for loading and elution buffers [31].

Before entering the analytical column for LC-MS analysis, samples need to be free from salt, large polypeptides and detergents. Large polypeptides can be easily removed by filtering the sample through a membrane cut-off filter. Detergents are commonly used for efficient protein extraction, but even small amounts of detergent in the sample disturb the LC-MS analysis (*see* **Note 3** about detergent removal). Instead of detergents, chaotropes can be used for protein extraction. The classical example is concentrated urea (8–10 molar), which can be removed from the sample together with salts. Desalting is then generally performed using the C_{18}-packed tips that trap the polypeptides while salts (and chaotropes) pass through. Trapped peptides are eluted by 50 % acetonitrile and can be further analyzed. This approach was used by us and is described below in detail. Alternatively, online desalting can be employed using trapping units in LC-MS system [32]. It is important to note, that these considerations are for in-solution digestion protocols, while in-gel based approaches do not require any additional cleanup. It is recommended to check the purity of samples before LC-MS analysis as described in **Note 4**.

Further, the choice of mass spectrometer is crucial for the result of analysis. Nevertheless, this is the most inflexible parameter due to the very high cost of this instrumentation. In this protocol we describe how sample preparation is performed to be compatible with the electrospray ionization used in LC-MS as well as detailed description of settings used for viral proteome analysis using the linear ion trap Fourier transform (LTQ-FTICR) and instrument. The strategies for proteome virus analysis of Adenovirus are summarized in Fig. 1.

3.1 Virus Production

1. Grow HeLa cells in MEM medium with spinner modification in a 37 °C incubator with 5 % CO_2. Start the cells in a small flask and transfer then all cells to a spinner flask. Use a magnetic stirrer to disperse the cells. Expand the culture by daily counting the cells and addition of more medium. Keep the cell density not higher than 1×10^6 cells/mL before dilution and not below 0.3×10^6 cells/mL after dilution.
2. Infect the cells with adenovirus when the total cell culture volume is ~4 L. Spin down cells in 225 mL graduated conical tubes. Spin at ~600 × *g*, 4 °C for 15 min. Remove the medium and save it for further use. Resuspend the infected cells in 400 mL medium for infection. Add adenovirus at a 5–10 FFU (fluorescent focus forming units) per cell. Incubate for 2–3 h at 37 °C with stirring. Then, add back 3,600 mL of saved medium for cell growth to a final volume of 4 L. Allow the virus to grow at 37 °C for 72 h before harvest.
3. Spin down infected cells in four 225 mL graduated conical tubes at ~600 × *g*, 4 °C for 5 min. Four cell pellets are obtained.

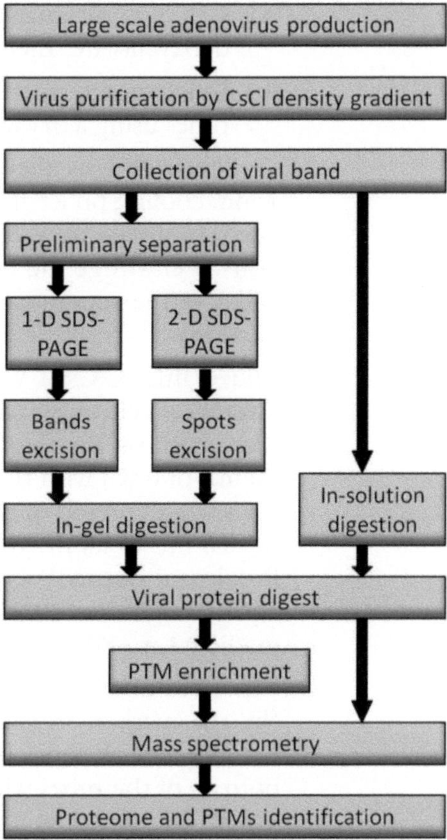

Fig. 1 Schematic illustration of different approaches for analysis of the total proteome and the posttranslational modifications in Adenovirus using mass spectrometry. After growing virus in host cells it is necessary to remove host cell proteins. Highly purified virus can be obtained by CsCl gradient centrifugation. Different alternatives are possible for fractionation of proteins, e.g., 1D and 2D SDS-PAGE before enzymatic cleavage of proteins into peptides. An alternative is to directly cleave proteins into peptides in solution. At the peptide stage, different enrichments for PTMs can be applied. The final detection of peptides is performed with mass spectrometry. After database search, protein identities and possible posttranslational modifications are reported. Manual inspection of the recorded MS/MS spectrum is required for verifying the position of a posttranslational modification in a peptide sequence

Remove as much supernatant as possible. Resuspend each pellet in 10 mL of 10 mM Tris–HCl, pH 7.9. Spin down at ~250 × *g*, 4 °C for 5 min. Resuspend again each pellet in 10 mL of 10 mM Tris–HCl pH 7.9 and freeze four aliquots at −80 °C until further use.

3.2 Virus Purification

1. Thaw the infected cell solutions at room temperature (we recommend using all four tubes from Subheading 3.1.3 and pooling them into one highly concentrated virus preparation at the end).

(a) Cell lysis, alternative 1: Add 1/10 volume of 5 % sodium deoxycholate. Incubate the cell solution on ice for 30 min.

(b) Cell lysis, alternative 2: Freeze–thaw the cell solution three times using a dry ice–ethanol bath and a 37 °C water bath.

For both lysis alternatives: Sonicate the cells for 15 s with 1 min cooling on ice in between. Repeat until the sample has a viscosity similar to water (approximately three times of sonication). Centrifuge the sample at ~2,000 × *g* at 4 °C for 10 min and save the supernatant. Pool processed cell lysates into two fractions.

2. Add 6 mL of CsCl with the density of 1.41 g/cm³ to two ultra-clear centrifuge tube. Mark the volume of this CsCl layer with a pen on the outside of the tube and gently overlay with 10 mL of CsCl with the density of 1.27 g/cm³. Then, add the virus lysates on top of the CsCl layers. (Adjust the virus volume to ~18 mL to fill the tube by diluting virus with storage buffer for virus if needed). Use a balance to adjust the two tubes to equal weight (< ±0.01 g). Ultracentrifugate (discontinuous gradient) at ~80,000 × *g* for 2 h at 4 °C.
3. The virus now forms a band between the two CsCl densities (by the marker on the tube) that can be seen by eye when holding a black paper behind the tube. Carefully puncture the bottom of the tubes with a sharp-edged tool and let the liquid slowly drip out. Use a waste tube to collect most of the solution that drips before the virus close band. Then change to a clean Eppendorf tube and collect the virus band.
4. Pool the two collected virus band to a fresh ultra-clear centrifugation tube and fill it up to ~34 mL with CsCl with the density of 1.34 g/cm³. Mix gently by pipetting up and down three to four times. Prepare a counter-balance tube only filled with CsCl, and as described in Subheading 3.2.2. Ultracentrifugate (equilibrium gradient) at ~80,000 × *g* for 18–22 h at 4 °C.
5. A virus band will now appear at the upper part of the tube and should be collected as described above (Subheading 3.2.3).
6. Remove the CsCl by dialyzing the virus against 10 mM Tris–HCl, pH 8.0 at 4 °C overnight using a Slide-A-Lyzer Cassette. Store the virus in 50–100 μL aliquots at −80 °C.
7. Check the purity of the virus after purification with CsCl gradient centrifugations using a 1D SDS-PAGE (refer to Subheading 3.3.2) stained with Coomassie. This is required to assure that no human proteins are left. Only distinct bands of expected proteins from the virus should be detected and if there is a new virus that is being studied, it is necessary to cut out bands and analyze with MS to get protein identity.

3.3 Preparation of Samples for Analysis of the Viral Proteome by Mass Spectrometry

3.3.1 In-Solution Based Approach

1. Lysis and protein concentration measurement: Lyse a small aliquot of virus by adding the reagents or aliquots of stock solutions directly to the virus solution (*see* Subheading 2.3.2). Sonicate 3 × 10 s at 15 micron. Use viral lysates to determine protein concentration in your virus batch. For later MS analysis it is necessary to estimate how much proteins are digested.
2. Lysis and digestion: Take out a desired amount of virus (concentration determined in **step 1**) (normally 10 μg will be enough for a number of replicates in total protein analysis and 50 μg will be suitable per phosphopeptide enrichment) and add reagents or aliquots of stock solutions and inhibitors directly to the virus solution (*see* Subheading 2.3.2). Sonicate 3 × 10 s at 15 micron. Prepare 45 mM DTT solution from stock and add 1/10 of the sample volume. Incubate 20 min at 60 °C. Prepare 100 mM IAA solution. Add 1/10 volume of IAA solution to the sample and incubate in the dark, at room temperature, shaking at 400 rpm. Add 0.3× lysis volume of 10× HEPES buffer and then dilute the sample to a volume corresponding to 4× lysis volume with Milli-Q water (Example: If 50 μL of lysates, the final volume will be 200 μL after dilution. Then 200 × 0.3 = 60 μL of 10× HEPES should be added and the volume of water to be added is 200 μL—50 μL (lysates volume)—5 μL (DTT solution)—5 μL (IAA solution)—60 μL (10× HEPES solution) = 80 μL). Add trypsin in a ratio 1:50 of trypsin–protein. Digest at 37 °C, shaking at 400 rpm, overnight. (For other enzymes, use conditions recommended by the supplier). Proteins are now cleaved into peptides. Acidify the sample by adding TFA to a final concentration of 1 %.
3. Sample filtration: Filter the sample through a 10 kDa MWCO ultrafiltration unit to remove uncleaved and large polypeptides that cannot be analyzed by LC-MS and instead could clog the system. Save the filtrate and check that there <10 % of loaded sample left in the upper part.
4. Sample desalting: Use for example the Stage-Tip technology to desalt the peptides [33]. Punch an Empore Disks C_{18} membrane twice with a G20 needle and transfer the pieces into the edge of a 200 μL pipette tip. Use a 1.5 mL Eppendorf tube and punch the lid so the prepared tip can fit therein. Load 50 μL 90 % acetonitrile, 0.1 % TFA solution and spin down. Use table centrifuge at a maximum speed of 1,500 × *g*. Load 50 μL of 0.1 % TFA and spin down, then load the sample in 50–100 μL portions and spin down. Wash with 20 μL 0.1 % TFA. Place the tip in a new Eppendorf tube and elute the peptides with 50 μL of 50 % acetonitrile in 0.1 % TFA Milli-Q water. Dry the sample down in a SpeedVac and store in −20 °C until MS analysis.

For troubleshooting, *see* **Note 3** about how to remove detergents, **Note 4** on how to check sample quality and **Note 5** for control of digestion efficiency.

3.3.2 (1D) SDS-PAGE Separation

1. Take out desired volume/amount of purified virus, e.g., 25 μL corresponding to 12–25 μg, and aim for a total volume of 100 μL. Add 20 μL of 5× Laemmli sample loading buffer, 10 μL of 45 mM DTT, and 35 μL of Milli-Q water (adjust the water volume depending on what volume of virus is used). Incubate at 95 °C for 5 min. Cool until room temperature and then add 10 μL of 100 mM IAA. The total volume is now 100 μL. Incubate 5 min in the dark at room temperature (if less than 100 μL of sample is needed, divide the volumes, or for concentrated samples, replace water with virus sample). The sample is now reduced and alkylated.
2. Analyze the sample on a 1D SDS-PAGE known to resolve all expected protein bands. Gels can be purchased pre-cast or produced in-house following generally available protocols. We recommend a 13 % gel to resolve and detect the proteins in the Ad2 virus capsid. Add the sample together with a low molecular weight marker and run the gel with settings suitable for your type of electrophoresis system and gel type. Stop the gel when the front migrates out to make sure that low molecular proteins are not lost.
3. Take out the gel and treat it with fixation solution. Incubate for 45 min to 1 h.
4. Mix methanol and Coomassie staining stock solution 1:4 and incubate overnight at room temperature. Destain the gel by washing several times with water.

3.3.3 (2D) SDS-PAGE Separation

1. Protein extraction using 2D electrophoresis: Take out a volume of virus solution corresponding to around 500 μg. A volume around 500 μL is recommended. Add reagents and aliquots of stock solutions of buffers and inhibitors directly to the virus (*see* Subheading 2.3.4). Adjust with Milli-Q water to the desired final concentrations. Shear the viral DNA by pipetting up and down with a needle (e.g., 21G, 0.8×40 mm) connected to a syringe. Transfer the lysates to an ultra-clear ultracentrifugation tube. Overlay and fill the tubes with mineral oil. Prepare a centrifuge tube for counterbalancing as described in Subheading 3.2.2, and load the tubes into a SW60Ti rotor. Use an ultracentrifuge at ~100,000×*g* for 30 min at 4 °C. Take out the sample carefully by using a pipette.
2. Loading of sample onto an IEF strip: Take out a sample volume corresponding to 30–100 μg of viral proteins lysed sample preparation solution for 2D electrophoresis. Add 0.625 μL of

1 M DTT solution and 0.625 μL IPG buffer (compatible with the choice of pH range in IPG strip) and adjust the volume to 125 μL with rehydration buffer.

3. Add the sample solution into the Immobiline Rehydration tray and gently place an Immobiline DryStrip of pH interval 4–7 of 3–11 NL onto the solution with gel-side down. Overlay the strip with DryStrip Cover Fluid and re-swell the strip for 10–20 h. *See* **Note 6** about Immobiline DryStrip and rehydration loading.
4. Move the strip to an Ettan IPGphor Isoelectric Focusing Unit and focus the proteins using the following program: (a) step the voltage to 300 V for 0.3 h, (b) Increase the voltage to 5,500 V by a gradient, for 0.3 h; and (c) Hold the voltage at 5,500 V until 22,000 Vh are reached. Take out the strip with a tweezer and store at −80 °C in a 15 mL Falcon tube with gel-side facing the internal until next step.
5. Place the strip in a 15 mL Falcon tube and add 5 mL equilibration buffer supplemented with DTT. Incubate at room temperature for 15 min with rotation.
6. Replace the solution with 5 mL equilibration buffer supplemented with IAA and incubate for another 15 min with rotation.
7. Place the strip with the anodic (+) end to the left and the gel side facing forward (towards you) into a 1D SDS-PAGE gel with one large well. Use a commercial 4–12 % acrylamide gel or 13 % acrylamide if the gel is prepared in house. Seal the gel using the agarose sealing solution.
8. Run second dimension as described for 1D SDS-PAGE, fix the gel and stain with Coomassie. Pick up protein spots of interest.

3.3.4 In-Gel Digestion (Suitable for Excised Pieces from Both 1D and 2D SDS-PAGE)

1. After appropriate destaining of the gel, excise the virus protein bands or spots of interest. Divide the bands/spots into small pieces of approximately 1 × 1 mm and add to an Eppendorf tube. Make sure to handle the gel on a clean surface. Use gloves, lab coat, hair cover, and clean tools, e.g., a scalpel or punch needle, to avoid contamination.
2. Destain the gel pieces by adding 100–200 μL 50 % acetonitrile, 50 % 100 mM NH_4HCO_3 solution. Incubate at room temperature in a shaker (400 rpm) for 20 min. Discard the liquid. Repeat this step until most Coomassie stain has disappeared, but for a maximum of two times. Add 100 % acetonitrile and incubate for 5 min. Discard the acetonitrile and dry the gel piece at 50 °C in for example a SpeedVac. The gel pieces will now become white and electrostatic.

3. Sample digestion: Dilute trypsin to a concentration of 25 ng/μL in 50 mM NH_4HCO_3 and add 3–10 μL per gel piece (depending on intensity) and keep on ice for 1 h. Add 20–30 μL 50 mM NH_4HCO_3 (depending on size of the gel piece) and incubate at 37 °C overnight.
4. Peptide extraction: Sonicate the gel pieces in the surrounding solution and collect the liquid. If the volume has dramatically decreased, add few more μL of 50 mM NH_4HCO_3 before sonication. Dry the liquid down in a SpeedVac and store the samples at −20 °C until analysis.

3.3.5 Enrichment of Phosphopeptides

The TiO_2 material is provided by the manufacturer packed in tips, Spin Tip (or can be produced in house [31]) and should be handled using a centrifuge. *See* **Note 7** for specific enrichment of tyrosine phosphorylated peptides.

1. Put the Spin Tip in a punched lid of an Eppendorf tube that will be used as waste tube. Check that the tip does not meet the bottom of the tube.
2. Add 20 μL of loading buffer A and then centrifuge at 3,000 × *g*, room temperature, for 2 min to condition the TiO_2 material.
3. Add 20 μL of loading buffer B and centrifuge 3,000 × *g*, room temperature, for 2 min to equilibrate the TiO_2 material. Remove the 40 μL collected in the waste tube.
4. To adsorb the peptides, dissolve desalted peptides in 100 μL loading buffer B and add this to the Spin Tip. Centrifuge at 1,000 × *g*, room temperature, for 10 min. Collect the solution that passes through the Spin Tip and put it back into the Spin Tip again. Centrifuge at 1,000 × *g*, room temperature, for 10 min. Remove the liquid collected in the waste tube.
5. Rinse the peptides by adding 20 μL of loading buffer B and centrifuge at 3,000 × *g*, room temperature, for 2 min. Rinse by again by adding 20 μL of loading buffer A and an additional centrifugation at 3,000 × *g*, room temperature, for 2 min.
6. Discard the waste tube and put the Spin Tip into a new Eppendorf tube. Elute the adsorbed peptides by adding 50 μL of elution buffer A and centrifuge at 1,000 × *g*, room temperature, for 5 min. Then add 50 μL of elution buffer B and centrifuge at 1,000 × *g*, room temperature, for 5 min.
7. Acidify the peptides by adding 100 μL of 20 % TFA. Confirm the acidic pH by pH-paper. Then, desalt the peptides according to the StageTip procedure described in Subheading 3.3.1. To check sample quality, *see* **Note 4**.

3.4 LC-MS Analysis

Here we present the protocol for using a 7-tesla LTQ-FT Ultra tandem mass spectrometer modified with a nano electrospray ion source (ProxeonBiosystems) for viral proteome LC-MS analysis. For users of other instruments, **Note 8** describes the main parameters of a mass spectrometer that need to be taken into account.

1. Set a 40-min linear gradient from 2 % to 45 % of buffer B at 200 nL/min on Agilent 1100 Nanoflow system equipped with homemade 15-cm fused silica emitter packed with Reprosil-Pur C18-AQ 3 μm resin (Dr. Maisch GmbH), and coupled on-line to the mass spectrometer.
2. Create a program for operating the mass spectrometer in positive ion mode that automatically switches between a high resolution (resolving power 50,000) survey mass spectrum in the FTMS cell and consecutive low resolution CID (collisionally induced dissociation) spectra of the 5 most abundant ions in the ion trap. Set normalized collision energy to 30 %. Set dynamic exclusion for 30 s for 5 peptides.
3. Dilute samples in water–TFA (1:0.005 v/v), load them into LC-MS system and acquire the data.
4. Convert the acquired data (.RAW-files) to Mascot generic format (MGF) files using appropriate software. Submit acquired MGF files to Mascot search engine (version 2.1.3, Matrix Science) by searching the Uniprot-Swissprot database with following settings:
   (a) Set “Viruses” taxonomy.
   (b) Set the enzyme specificity (i.e., trypsin) and number of allowed missed cleavages (we used two in our experiments).
   (c) Set mass deviation for precursor and fragment ions. For our runs with LTQ-FTICR they were set to 0.02 Da and 0.7 Da correspondingly.
   (d) Set “Carbamidomethylation” as a fixed modification.
   (e) Set variable modifications to deamidation (N, Q) and oxidation (M). Then set up to for extra modifications if PTM identification is required, for instance, extra phosphorylation (S,T), phosphorylation (Y), methylation (C-terminus), and methylation (D, E) can be set (*see* also **Note 8**).
   (f) Choose instrument setting (i.e., “ESI-FTICR”)
5. The search results will present the identified proteins, peptides, and suggested PTMs. The identifications have a certain probability specified in the result file, and newly discovered PTMs must be validated. We recommend manual interpretation of MS/MS spectra [34] for verification. Examples of verified PTM spectra are presented in Fig. 2. As a reference database containing already discovered viral PTMs we recommend online virPTM database [35] (http://virptm.hms.harvard.edu/).

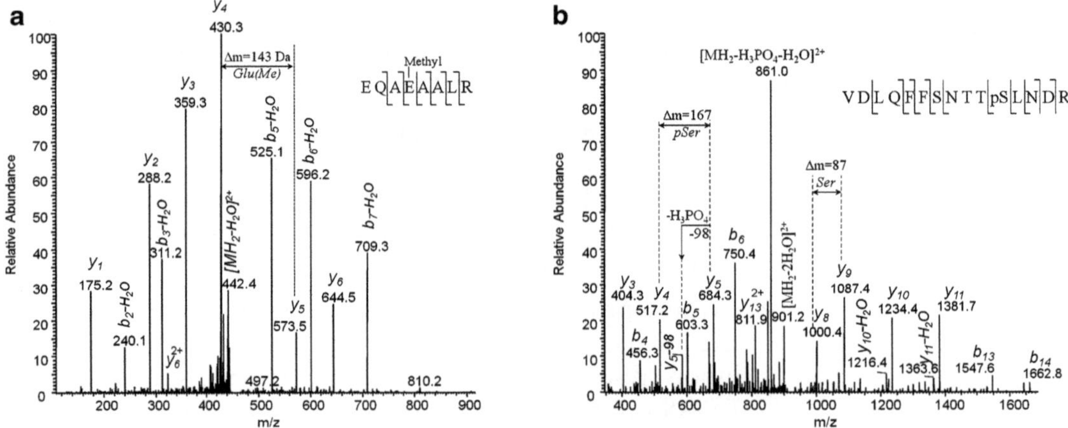

Fig. 2 Manual verification of posttranslational modifications, identified by Mascot. (**a**) Methylation of glutamic acid. Mascot reported a peptide sequence with low mass error (<5 ppm) originating from the viral proteome and a methylation on aspartic or glutamic acid (Methylation (D, E)) as modification. The theoretical backbone cleavage was clearly observed (normally with signal/noise (S/N) >3). In this example ion y_5 has high S/N ratio and identifies methylated glutamic acid in position 4. Mass difference between y_4 and y_5 ions equals 143 Da (Glutamic acid (Glu) +14, i.e., CH_2 for methylation). Position 1 lacks methylation, that is confirmed by the loss of H_2O molecule from carboxylic group of b_2 ion ($COOCH_3$ cannot lose H_2O molecule), as well as by b_2 ion mass of 240.1 Da (Glu-H_2O+Gln+1). (**b**) Phosphorylation. Similarly to methylation, pairs of y_4 and y_5 ions with S/N > 3 and mass difference of 167 Da identify phosphoserine in position 11. Observed neutral loss of 98 Da (phosphoric acid) from y_5 ion confirms that serine 11 is phosphorylated. Additional evidence is a clear mass difference of 87 Da (unmodified serine) between y_8 and y_9 ions, i.e., serine 7 is not phosphorylated. Exact mass of this peptide indicates the presence of only one phosphorylation. Therefore the phosphorylation at serine 11 is confirmed by manual verification

4 Notes

1. All reagents to be used for sample preparation for MS analysis should have purity specified for LC-MS application and water used should be so-called Milli-Q water. Sterile solutions are required during the virus production steps. We strongly recommend processing all samples from the designed study using chemicals and plastic material from the same batch in order to avoid "batch effects" [36].
2. The detectability of different proteins in a virus particle will vary depending on a number of parameters. In Table 1 we have summarized the characteristics and the results obtained for analysis of Adenovirus type 2 using mass spectrometry. The number of copies of a certain protein per virion will reflect the abundance of peptides originated from this protein in the sample. For example in adenovirus type 2 the hexon protein is dominating in the virus particle, and peptides of the hexon protein

will compete during analysis with peptides from other proteins. Surprisingly, separation of adenovirus sample on 1D SDS-PAGE, excising bands, in-gel digestion, and pooling of all protein bands except for the hexon was not beneficial compared to direct LC-MS analysis of the whole virion sample digested in-solution. We believe, that the high capacity of the LC-MS system in comparison to losses in the in-gel digestion approach makes the in-solution digestion superior. However, if analyzing the protein bands one by one, higher sequence coverage for each protein is expected. Further, the sequence coverage of a protein depends on the amino acid sequence of the proteins in relation to what protease that is used for digesting the protein into peptides. For example, the pV and pVII proteins are very rich in arginine and lysine residues, which are the cleavage sites for trypsin. Then, very small peptides that cannot be detected with the MS instrument are expected, which can explain the low sequence coverage of those proteins. On the other hand, proteins containing few arginines and lysines will produce long peptides that might escape detection. In the same way, the chances of detecting PTMs will be influenced by the choice of protease.

3. In order to remove the detergents offline or online-coupled trap columns can be used, for example, SCX (for anionic detergents like SDS) or SAX (for cationic detergents). The sample is loaded on such trap column and detergents remain in it, while peptides pass through for further analysis. Unfortunately, many peptides are also trapped by such columns, which cause sample loss. Use of "MS-friendly" nonionic detergents might be beneficial in this case. β-octylglucoside (OBG) and RapiGest are the best choices. OBG has significantly lower impact on LC-MS runs, and can be easily removed by ethylacetate extraction [37]. RapiGest is destroyed by adding 20 % TFA followed by centrifugation, and its degradation products do not affect LC-MS runs [38].

4. Quality check of sample before LC-MS: An easy way to find out if there are contaminants that influence the results in ESI-LC-MS in the sample, e.g., detergents, is to use MALDI-TOF MS analysis. Follow a general protocol for sample preparation and analysis [39] and load about 0.5 μL of your sample.

5. Control of in-solution digestion efficiency. If having low recovery of peptides after digestion it is wise to check if the digestion has been complete. Follow the description for 1D SDS-PAGE and load a small aliquot of sample before and after digestion. Fix and stain the gel with Coomassie. Check that there are no uncleaved bands in the lane of the digested sample.

6. For 2D SDS-PAGE we have tried different loading techniques and pH intervals and found that the rehydration loading

worked best for Adenovirus type 2. The very basic proteins are difficult to identify even when using basic Immobiline Dry strips pH 7–11. Therefore we recommend pH 4–7 or pH 3–11 strips. The 7 cm strips have sufficient capacity for the Adenovirus proteome since they only consist of a dozen of proteins with a maximum of ten isoforms.

7. For specific enrichment of tyrosine phosphorylation, a larger amount of virus, i.e., around 1 mg is recommended to be incubated with anti-tyrosine phosphorylated antibody as described by Bergström Lind et al. [23].

8. As mentioned before, MS instrumentation is the most inflexible parameter due to the high cost of the equipment. We describe the approach we have used with our LTQ-FTICR mass spectrometer, but one can use alternative instruments. Here we list the most important settings and parameters that need to be taken into account during LC-MS experiments.

   (a) Resolution. This is a very important parameter for obtaining reliable protein and, especially, PTM identifications. With a higher resolution a higher precision of mass determination is possible. Better confidence is then achieved when matching the theoretical mass of the proposed sequence to the experimental value. Orbitrap and high-resolution Q-TOF instruments can be alternatives to LTQ-FTICR. The most crucial is to record the survey scan with high resolution, while fragmentation spectra can be less resolved [23].

   (b) Acquisition rate. This parameter is very important for highly complex samples, while viruses are normally composed of a limited number of proteins, e.g., Ad2 consist of about a dozen. Therefore it is not critical to detect a maximum number of peptides. More important is to record high quality spectra for a limited set. Ion traps or conventional Q-TOFs are very fast mass spectrometers with low resolution and therefore less useful. If they are the only available choice, the user must specify as high resolution as possible sacrificing scanning speed.

   (c) Fragmentation. Overall, CID provides very good results for proteome profiling and is recommended for most mass spectrometers. Orbitrap instruments offer a special HCD (Higher-energy collisional dissociation) fragmentation possibility, which is more effective than CID. HCD was also shown to be the best fragmentation technique for identification of phosphorylated peptides. Available on many instruments, ETD (electron transfer dissociation) or ECD (electron capture dissociation) fragmentation is very useful for localization of labile PTMs, like glycosylations and phosphorylations on serine and threonine. They are commonly used as a complement to CID or HCD spectra.

(d) Data analysis. Mascot is the most common but not the only search engine that can be used for viral proteome analysis. Alternative engines are for example SEQUEST and X!Tandem. The general rule is that no search engine works properly if more than six variable modifications are set. Consequently, to search for more than four PTMs, it is necessary to perform several consecutive searches using different sets of variable PTMs.

Acknowledgement

This work was supported by the P.O. Zetterling Foundation (SBL), the Swedish Cancer Society (SBL), and the Kjell and Märta Beijer Foundation (UP). The authors wish to thank Lioudmila Elfineh for valuable discussions.

References

1. Flint SJ (2001) Adenovirus. Encyclopedia of life sciences. Wiley. http://www.els.net
2. Russell WC (2000) Update on adenovirus and its vectors. J Gen Virol 81:2573–2604
3. Stewart PL, Fuller SD, Burnett RM (1993) Difference imaging of adenovirus: bridging the resolution gap between X-ray crystallography and electron microscopy. EMBO J 12: 2589–2599
4. van Oostrum J, Burnett RM (1985) Molecular composition of the adenovirus type 2 virion. J Virol 56:439–448
5. Fuschiotti P, Schoehn G, Fender P, Fabry CM, Hewat EA, Chroboczek J et al (2006) Structure of the dodecahedral penton particle from human adenovirus type 3. J Mol Biol 356:510–520
6. Rux JJ, Kuser PR, Burnett RM (2003) Structural and phylogenetic analysis of adenovirus hexons by use of high-resolution x-ray crystallographic, molecular modeling, and sequence-based methods. J Virol 77: 9553–9566
7. Zubieta C, Schoehn G, Chroboczek J, Cusack S (2005) The structure of the human adenovirus 2 penton. Mol Cell 17:121–135
8. Liu H, Jin L, Koh SB, Atanasov I, Schein S, Wu L et al (2010) Atomic structure of human adenovirus by cryo-EM reveals interactions among protein networks. Science 329:1038–1043
9. Reddy VS, Natchiar SK, Stewart PL, Nemerow GR (2010) Crystal structure of human adenovirus at 3.5 A resolution. Science 329: 1071–1075
10. Washburn MP, Wolters D, Yates JR 3rd (2001) Large-scale analysis of the yeast proteome by multidimensional protein identification technology. Nat Biotechnol 19:242–247
11. Johnson RS, Davis MT, Taylor JA, Patterson SD (2005) Informatics for protein identification by mass spectrometry. Methods 35: 223–236
12. Jensen ON (2006) Interpreting the protein language using proteomics. Nat Rev Mol Cell Biol 7:391–403
13. Thingholm TE, Jensen ON, Larsen MR (2009) Analytical strategies for phosphoproteomics. Proteomics 9:1451–1468
14. Bodenmiller B, Mueller LN, Mueller M, Domon B, Aebersold R (2007) Reproducible isolation of distinct, overlapping segments of the phosphoproteome. Nat Methods 4: 231–237
15. Wisniewski JR, Zougman A, Kruger S, Mann M (2007) Mass spectrometric mapping of linker histone H1 variants reveals multiple acetylations, methylations, and phosphorylation as well as differences between cell culture and tissue. Mol Cell Proteomics 6:72–87
16. Bergstrom Lind S, Artemenko KA, Elfineh L, Zhao Y, Bergquist J, Pettersson U (2012) The phosphoproteome of the adenovirus type 2 virion. Virology 433:253–261

17. Akusjarvi G, Philipson L, Pettersson U (1978) A protein kinase associated with adenovirus type 2. Virology 87:276–286
18. Axelrod N (1978) Phosphoproteins of adenovirus 2. Virology 87:366–383
19. Blair GE, Russell WC (1978) Identification of a protein kinase activity associated with human adenoviruses. Virology 86:157–166
20. Russell WC, Blair GE (1977) Polypeptide phosphorylation in adenovirus-infected cells. J Gen Virol 34:19–35
21. Tsuzuki J, Luftig RB (1983) The adenovirus type 5 capsid protein IIIa is phosphorylated during an early stage of infection of HeLa cells. Virology 129:529–533
22. Pettersson U, Sambrook J (1973) Amount of viral DNA in the genome of cells transformed by adenovirus type 2. J Mol Biol 73:125–130
23. Bergstrom Lind S, Artemenko KA, Pettersson U (2012) A strategy for identification of protein tyrosine phosphorylation. Methods 56:275–283
24. Qoronfleh MW, Ren L, Emery D, Perr M, Kaboord B (2003) Use of immunomatrix methods to improve protein-protein interaction detection. J Biomed Biotechnol 2003:291–298
25. Choudhary C, Kumar C, Gnad F, Nielsen ML, Rehman M, Walther TC et al (2009) Lysine acetylation targets protein complexes and co-regulates major cellular functions. Science 325:834–840
26. Na CH, Peng J (2012) Analysis of ubiquitinated proteome by quantitative mass spectrometry. Methods Mol Biol 893:417–429
27. McDonald CA, Yang JY, Marathe V, Yen TY, Macher BA (2009) Combining results from lectin affinity chromatography and glycocapture approaches substantially improves the coverage of the glycoproteome. Mol Cell Proteomics 8:287–301
28. Dunn JD, Reid GE, Bruening ML (2010) Techniques for phosphopeptide enrichment prior to analysis by mass spectrometry. Mass Spectrom Rev 29:29–54
29. Zhou H, Watts JD, Aebersold R (2001) A systematic approach to the analysis of protein phosphorylation. Nat Biotechnol 19:375–378
30. Goshe MB, Conrads TP, Panisko EA, Angell NH, Veenstra TD, Smith RD (2001) Phosphoprotein isotope-coded affinity tag approach for isolating and quantitating phosphopeptides in proteome-wide analyses. Anal Chem 73:2578–2586
31. Thingholm TE, Jorgensen TJ, Jensen ON, Larsen MR (2006) Highly selective enrichment of phosphorylated peptides using titanium dioxide. Nat Protoc 1:1929–1935
32. Zhang Z, Smith DL, Smith JB (2001) Multiple separations facilitate identification of protein variants by mass spectrometry. Proteomics 1:1001–1009
33. Rappsilber J, Mann M, Ishihama Y (2007) Protocol for micro-purification, enrichment, pre-fractionation and storage of peptides for proteomics using StageTips. Nat Protoc 2:1896–1906
34. Kinter M, Sherman NE (2000) Protein sequencing and identification using tandem mass spectrometry. John Wiley & Sons, Inc., New York. ISBN-13: 978-0-471-32249-8
35. Schwartz D, Church GM (2010) Collection and motif-based prediction of phosphorylation sites in human viruses. Sci Signal 3:rs2
36. Gregori J, Villarreal L, Mendez O, Sanchez A, Baselga J, Villanueva J (2012) Batch effects correction improves the sensitivity of significance tests in spectral counting-based comparative discovery proteomics. J Proteomics 75:3938–3951
37. Yeung YG, Nieves E, Angeletti RH, Stanley ER (2008) Removal of detergents from protein digests for mass spectrometry analysis. Anal Biochem 382:135–137
38. Yu YQ, Gilar M, Lee PJ, Bouvier ES, Gebler JC (2003) Enzyme-friendly, mass spectrometry-compatible surfactant for in-solution enzymatic digestion of proteins. Anal Chem 75:6023–6028
39. Jiménez CR, Huang L, Qiu Y, Burlingame AL (1998) Sample preparation for MALDI mass analysis of peptides and proteins. Curr Protoc Protein Sci 16.13.11–16.13.16

Chapter 3

Capsid Modification Strategies for Detargeting Adenoviral Vectors

Alan L. Parker, Angela C. Bradshaw, Raul Alba, Stuart A. Nicklin, and Andrew H. Baker

Abstract

Adenoviral vectors hold immense potential for a wide variety of gene therapy based applications; however, their efficacy and toxicity is dictated by "off target" interactions that preclude cell specific targeting to sites of disease. A number of "off target" interactions have been described in the literature that occur between the three major capsid proteins (hexon, penton, and fiber) and components of the circulatory system, including cells such as erythrocytes, white blood cells, and platelets, as well as circulatory proteins including complement proteins, coagulation factors, von Willebrand Factor, p-selectin as well as neutralizing antibodies. Thus, to improve efficacious targeting to sites of disease and limit nonspecific uptake of virus to non-target tissues, specifically the liver and the spleen, it is necessary to develop suitable strategies for genetically modifying the capsid proteins to preclude these interactions. To this end we have developed versatile systems based on homologous recombination for modification of each of the major capsid proteins, which are described herein.

Key words Adenovirus, Replication deficient, Gene therapy, Hexon, Penton base, Fiber, Capsid modification, Homologous recombination

1 Introduction

To date, some 57 different serotypes of human adenovirus have been described, divided into seven separate species A–G on the basis of their sequence alignments, hemagglutination patterns, and receptor usage (reviewed in [1]). Clinically, infections caused by adenoviruses tend to be self limiting, resulting in respiratory diseases (primarily species B and C), conjunctivitis (species B and D) and gastroenteritis (species F and G). Due to their remarkable capacity to transduce a diverse variety of cells both in vitro and in vivo, they are useful tools experimentally for over expression of transgenes. Furthermore, their progression to the clinic for therapeutic purposes has been rapid, and they are now the most commonly

Miguel Chillón and Assumpció Bosch (eds.), *Adenovirus: Methods and Protocols*, Methods in Molecular Biology, vol. 1089, DOI 10.1007/978-1-62703-679-5_3, © Springer Science+Business Media, LLC 2014

deployed clinical gene therapy vehicle, accounting for almost 25 % of ongoing clinical trials (http://www.abedia.com/wiley/vectors.php). However, their efficacy in clinical trials have been hindered by the host inflammatory response against the vector that results primarily from accumulation of the vector within non-target cell types and organs, primarily in the liver and spleen, resulting in significant dose limiting toxicities associated with the vector.

Both clinically and experimentally, the most studied and best understood serotype remains the species C adenovirus 5 (Ad5). In vitro, and for certain local applications in vivo (e.g., following intramuscular injection), Ad5 engages the Coxsackie and Adenovirus Receptor (CAR) as its primary attachment receptor [2], via interaction with specific key amino acids at positions within the globular fiber knob domain AB loop (Ser408 and Pro409), DG loop (Tyr477), and β strand F (Leu485) [3]. Where the virus is introduced systemically, it has been suggested the expression of CAR within platelets and red blood cells can cause "off target", toxicity inducing interactions, resulting in hemagglutination and thrombocytopenia respectively [4–6]. Thus, for systemic application of Ad5, genetic modulation of the fiber protein to ablate CAR mediated cell association may be advantageous in preventing dose limiting toxicities associated with platelet and red blood cell depletion (plus potential downstream effects on deposition within the liver). Following CAR engagement, the classical in vitro model of Ad5 cellular infection suggests that the virus then engages with αvβ3 and αvβ5 integrins at the cell surface to mediate cellular internalization, via Arg-Gly-Asp motifs in the penton base protein (positions 340–342) which stimulates internalization into clathrin coated vesicles [7]. However, for intravascular applications, recent studies have attributed a new and important role of this integrin–penton base interaction in mediating uptake of virus to the spleen with a corresponding induction of number of cytokines that promote a robust induction of innate immune responses [8, 9]. Thus it is clear that the genetic modulation of this interaction could profoundly improve both the safety and dose limiting toxicity of the virus for systemic applications. Further still, recent evidence has confirmed a compelling new role for the hexon protein in mediating cellular uptake via recruitment of blood clotting factors which coat the virus and "bridge" the virus: protein complex to highly sulfated heparan sulfate proteoglycan (HSPG) receptors expressed abundantly in liver hepatocytes [10–14]. Sequestration within liver hepatocytes not only serves to deplete the pool of vector available for therapeutic purposes but also further stimulates anti-vector immunity, and thus dose limiting toxicities. The hexon–factor (F) X interaction has been carefully modelled and critical interacting motifs located within and distal to the 5th and 7th hypervariable regions (HVRs) have been identified and genetically modified to preclude FX binding and hepatic transduction of the

Ad5 vector [15–17]. Gross modulation of the hexon hypervariable regions may also be advantageous in terms of evading preexisting anti Ad5 immunity, since 30–90 % (depending on ethnicity) of individuals have previously been exposed to Ad5 and thus present neutralizing antibodies (NAbs) against the vector [18–20]. The majority of studies have implicated the hexon HVRs as the major site of immunological recognition [20], at least following intramuscular challenge (the repertoire of immunological recognition following natural infection of airway epithelia may differ significantly [10, 21]). Therefore, applications which entail repeated administration of Ad5 will require the generation of genetically modulated viruses that can evade the host acquired immune response against the vector [20].

In addition to the well-documented interactions overviewed above, a number of other dose limiting interactions between virus and host proteins have been partially described in the literature that may preclude efficacious delivery via the systemic route. These interactions include (but are not limited to) interactions with complement proteins (C3, C1q, C4BP) [4, 22, 23], von Willebrand Factor, p-selectin [6], lactoferrin (produced locally at sites of inflammation [24]), and other proteins involved in the blood clotting cascade (FVII, FIX, Protein C, FX) [10]. It is therefore clear that the production of safer and more efficacious generations of adenoviral vectors for systemic gene therapy applications will necessitate protocols that permit the rapid genetic modification of the Ad5 capsid to prevent such dose limiting interactions and toxicities.

In a previous book [25] we have described detailed procedures for the production and purification of first generation Ad5 vectors, including methods for producing adenoviruses with point mutations (to ablate CAR interactions) or peptide insertions (for retargeting strategies) in the fiber protein, as well as point mutations within the hexon protein (to ablate FX interactions). Here we update and extend upon the protocols previously published to include some new quality control procedures for validating viral integrity (silver staining and use of NanoSight) as well as describing methods for ablating the penton base–integrin interaction, and exchanging the hexon hypervariable regions from less seroprevalent adenoviral species into Ad5, to reduce recognition by preexisting anti-hexon neutralizing antibodies and (potentially) to ablate interactions with blood clotting factors. Our strategies are based on "AdEasy-like" systems, necessitating the design of 4–8 kb shuttle plasmids containing the relevant gene, which are appropriate for genetic manipulation, flanked by ~500–1,500 bp flanking regions of homologous Ad5 sequence to permit homologous recombination with the backbone Ad5 genome cosmid to generate a full Ad genome for propagation of genetically modulated adenoviruses (overviewed in Fig. 1).

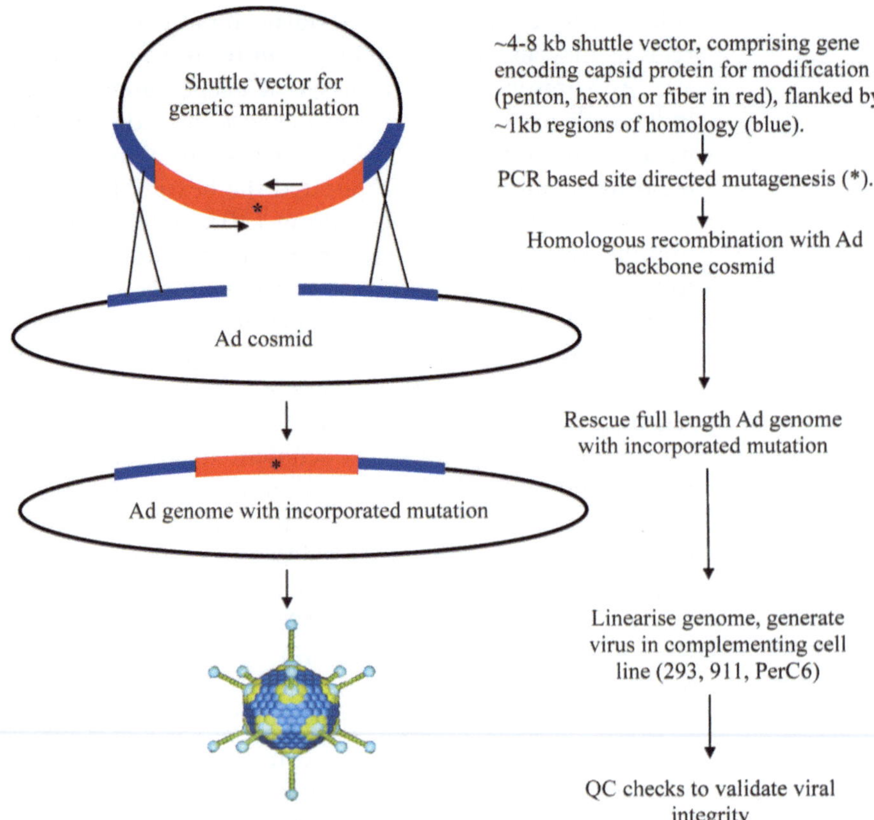

Fig. 1 Overview of strategy to produce capsid modified adenoviral vectors

Ultimately, these vectors should enable the production of safer, more efficacious vectors for in vivo applications whose tropism can be tailored towards individual applications.

2 Materials

2.1 Generation of Adenoviral Vectors with Altered Penton Base and Hexon Proteins

2.1.1 General Molecular Biology Reagents

1. Proof Reading *Taq* polymerase (Herculase II Fusion DNA Polymerase, #600677-41, Stratagene, CA, USA).
2. Rapid DNA Ligation kit (#11635379001, Roche, Mannheim, Germany).
3. StrataClone Blunt PCR cloning kit (Stratagene, #240207, CA, USA).
4. Annealing buffer: 10 mM Tris–HCl, pH 7.5–8.0, 60 mM NaCl, 1 mM EDTA.
5. DH5αTOP10 competent bacteria.
6. 1 kB and 100 bp molecular weight ladders.

2.1.2 Generation of an Ad5 Penton Base Shuttle Plasmid

We use the following oligonucleotides flanking the penton base gene and homologous regions. PCR of this region yields a 4.6 kb fragment containing the entire penton base sequence, flanked by 1.9 Kb (left) and 1 Kb (right) regions of homology from the Ad5 genome.

1. Oligonucleotide 1 (Forward): 5′ ATATGACGAGGACGAT GAGTACG 3′.
2. Oligonucleotide 2 (Reverse) 5′ CGCCGTACACCTCATCA TACAC 3′.
3. Oligonucleotides for sequencing (1 every 400 nucleotides).

2.1.3 Incorporation of Integrin Binding Mutations

We use the following oligonucleotides to introduce an aspartic acid (D) to glutamic acid (E) mutation (5′ CGCGGCGAC 3′>5′ CGCGGCGAA 3′) within the penton base RGD sequence by site directed mutagenesis of the 1.3 Kb *SexA*I-*Asc*I fragment of the penton shuttle vector.

1. Oligonucleotide 3 (penton forward) 5′ ACCCGTGTGTACC TGGTGGACAACAA 3′.
2. Oligonucleotide 4 (penton reverse) 5′ TTTCACTGACGGT GGTGATGGTGGG 3′.

   Primers for introducing the D>E mutation (Underlined sequences correspond to the mutation site.).
3. Oligonucleotide 5 (Forward): 5′ ATTCGCGGC**GAA**ACCT TTGCCTG 3′.
4. Oligonucleotide 6 (Reverse): 5′ GGCAAAGGT**TTC**GCCGCG AATGG 3′.
5. Oligonucleotides for sequencing (1 every 400 nucleotides).

2.1.4 Generation of Ad5 Hexon Shuttle Plasmid

We have previously described [25] the strategy we have used for generating a hexon shuttle plasmid, where the hexon gene is amplified to incorporate an additional flanking 1 Kb DNA from the unique *BamH*I and *AsiS*I restriction sites present in the Ad5 genome, using the oligonucleotides 5′ CTCCTTATTCCACTGATCGCC 3′ (forward) and 5′ATCTGATCTCCGACAAGAGCG 3′ (reverse). Within the hexon gene is a unique *Nde*I site, which facilitates cloning strategies.

2.1.5 Swapping HVR7 from Ad26 into the Ad5 Hexon Gene

Oligonucleotides flanking the region of interest (Ad5 region between *Nde*I and *BamH*I).

1. Oligonucleotide 7 (Ad5 hexon forward Nde I) 5′ CCAATGAAACCATGTTACGG 3′.
2. Oligonucleotide 8 (Ad5 hexon reverse BamH I) 5′ TCGTCCATGGGATCCACC 3′.

3. Oligonucleotide 9 (Ad5 Ad26 HVR7 fusion forward-1) 5′ TGCTTTCCACTGAATGGCACTGGAACCAATTCC 3′.
4. Oligonucleotide 10 (Ad5 Ad26 HVR7 fusion reverse-1) 5′ AGTGCCATTCAGTGGAAAGCAGTAATTTGG 3′.
5. Oligonucleotide 11 (Ad5 Ad26 HVR7 fusion forward-2) 5′ ACAAAACCAAATAAGAGTTGGAAATAATTTTGCC 3′.
6. Oligonucleotide 12 (Ad5 Ad26 HVR7 fusion reverse-2) 5′ AACTCTTATTTGGTTTTGTCTAGAAATTGCATCG 3′.
7. Oligonucleotides for sequencing (1 every 400 nucleotides).

2.1.6 Swapping HVR5 from Ad26 into the Ad5 Hexon Gene

1. Oligonucleotide 7 (Ad5 hexon forward *Nde* I) 5′ CCAATGAAACCATGTTACGG 3′.
2. Oligonucleotide 8 (Ad5 hexon reverse *BamH* I) 5′ TCGTCCATGGGATCCACC 3′.
3. Oligonucleotide 13 (Ad5 Ad26 HVR5 fusion forward-1) 5′ CAATTTTTCGACGTCCCTGGCGGAAGTCC 3′.
4. Oligonucleotide 14 (Ad5 Ad26 HVR5 fusion reverse-1) 5′ AGGGACGTCGAAAAATTGCATTTCCAC 3′.
5. Oligonucleotide 15 (Ad5 Ad26 HVR5 fusion forward-2) 5′ AATACAAACCTAAAGTGGTATTGTACAGTG 3′.
6. Oligonucleotide 16 (Ad5 Ad26 HVR5 fusion reverse-2) 5′ CCACTTTAGGTTTGTATTCTTCCCCACTACC 3′.
7. Oligonucleotides for sequencing (1 every 400 nucleotides).

2.2 Additional QC Checks for Adenoviral Integrity Validation

2.2.1 Silver Staining

Silver staining allows high resolution and rapid detection of both core and capsid proteins within an adenovirus preparation, and has the advantage of much greater sensitivity compared to Coomassie staining. We have found that this procedure can often inform of capsid packaging deficiencies, where specific genetic alterations of the capsid prove incompatible with assembly. We recommend using the Pierce Silver Stain Kit, which gives good, reproducible band staining.

1. Gel Loading Buffer II (Life Technologies).
2. Denatured Adenovirus—typically we boil 5×10^{10} viral particles for 10 min before subjecting to electrophoresis on a 9 % SDS polyacrylamide gel.
3. Silver staining kit—Pierce Silver Stain Kit (Pierce, part # 24612).

2.2.2 Adenovirus Concentration and Monodispersity as Monitored Using NanoSight™

We have recently begun employing NanoSight™ technology within our lab for direct imaging of viral particles to establish viral prep monodispersity and concentration (for details *see* http://www.nanosight.co.uk). Our NanoSight™ is the LM14 model.

This laser-based nanoparticle tracking system is allows nanoscale particles such as viruses and virus aggregates to be directly and individually visualized in liquids in real-time, which allows a high-resolution determination of particle size and distribution. The technique is rapid and accurate (we note highly reproducible particle titers when directly compared to microBCA determination of protein levels). The technology measures simultaneously and directly measuring the diffusion coefficient of individual particles, using Nanoparticle Tracking Analysis (NTA) to automatically count and size the viruses and aggregates in a sample.

The NanoSight™ system works optimally within the range of 10^6 to 10^{10} viral particles per ml, and is ideally suited to nanoparticles 30–1,500 nm in size, and thus is ideal for detection of adenovirus (~90 nm). The system would not be suitable, however, for AAV preps, due to their smaller size. The tracking technology is based on speed of movement of nanoparticles under Brownian motion since particles of different sizes will scatter light differently. We have found that in addition to basic QC checks, the system can also be useful for studying interactions of adenoviral particles with human proteins in solution, especially to study whether such interactions can result in particle aggregation. The system can also be purchased with fluorescence filters in order detect and quantify fluorescently labelled particles.

3 Methods

3.1 Generation of Ad5 Vectors with Mutated Penton and Hexon Proteins

3.1.1 Generation of an Ad5 Penton Base Shuttle Plasmid

To facilitate cloning strategies, we design and use shuttle plasmids, typically 4–8 Kb in size encoding only a fragment of the Ad genome that are (due to their reduced size relative to the Ad genome) more amenable to manipulation of individual genes within the Ad5 genome. We aim to construct shuttle plasmids that contain the gene for manipulation and a minimum of 500 bp (ideally 1–3 Kb) of DNA sequence flanking this region to facilitate homologous recombination. For such strategies, it is critical to identify a unique restriction endonuclease site in pAdEasy-1, which is present in the target viral gene. For penton base, *Pme*I is unique in the Ad5 genome and therefore linearizes pAdEasy-1 in the penton base gene to allow homologous recombination within this specific region.

1. First use oligonucleotides 1 and 2 to PCR amplify a 4.6 Kb fragment from the AdEasy genome containing the entire penton sequence, flanked by 1.9 Kb (left) and 1 Kb (right) regions of homology from the Ad5 genome.
2. Confirm the correct size of the generated DNA fragment via agarose gel electrophoresis.

3. Purify the DNA band using a commercially available gel extraction kit (*see* **Note 1**).
4. Clone the purified PCR product into a plasmid, which does not contain the restriction sites that will be used in future cloning steps. For cloning steps, we recommend the use of Strataclone Blunt PCR cloning kit.
5. Sequence the entire fragment to confirm it does not contain any mutations.

The penton shuttle plasmid is ready for genetic manipulation to incorporate specific modifications.

3.1.2 Incorporation of Point Mutations into the Penton Gene

In order to introduce the relevant DNA point mutation (5′ GAC 3′—5′ GAA 3′) to encode a RG*D*-RG*E* mutation within the Ad5 penton base protein, that efficiently ablates interactions with αvβ3/5 integrins, we perform PCR based site-directed mutagenesis.

1. Using the penton base shuttle plasmid as the backbone, perform separate PCR reactions using oligonucleotide 3 and 6, and oligonucleotides 4 and 5.
2. Gel purify resultant fragments using commercially available kits.
3. Using equimolar amounts of the purified PCR products, amplify the full 1.3 Kb *Sex*AI-*Asc*I fragment, and sub clone back into the penton shuttle vector.
4. Sequence the insert to confirm presence of the correct modification, and check no other mutations have been introduced.
5. Linearize the penton shuttle vector using *Eco*RI, and the Ad5 backbone cosmid with *Pme*I.
6. Perform homologous recombination to rescue the full Ad genome incorporating the desired mutation in the penton base protein, using BJ5183 electroporation competent cells, and purify genome using commercially available kits.
7. Rescue recombinant adenoviral vectors using permissive cells (e.g. 293 cells) as described previously (*see* **Notes 2** and **3**).

3.1.3 Generation of an Ad5 Hexon Shuttle Plasmid

The adenovirus hexon protein, the most abundant of the capsid proteins, once considered to have only a passive, structural role in virus structure and function, has been of renewed and significant interest in recent years as roles have emerged for both liver targeting, through interplay with blood coagulation factors, and immunogenicity, through the interaction with neutralizing antibodies. Therefore, strategies for modulation of the exposed hypervariable regions (HVRs) are likely to be important both for reducing the host mediated innate anti viral response, as well as enabling rational targeting strategies through elimination of interactions with blood clotting factors, most importantly Factor (F) X. Here we overview the strategies we have used to replace the 5th and the 7th HVRs from Ad5 with those from the species D, low seroprevalence, non-FX interacting Ad26.

To genetically manipulate the hexon protein, we developed a hexon shuttle plasmid suitable for homologous recombination strategies with the full lengh Ad genome. We previously described the protocol we have used to generate a hexon shuttle plasmid using the primers 5′ CTCCTTATTCCACTGATCGCC 3′ and 5′ ATCTGATCTCCGACAAGAGCG 3′ [25]. This results in the amplification of the hexon gene incorporating an additional flanking 1 Kb DNA from the unique *BamHI* and *AsiSI* restriction sites present in the Ad5 genome. We used this shuttle plasmid for our strategies for swapping hexon HVR regions from Ad26.

3.1.4 Generating Ad5 Containing HVR 7 from Ad26

To generate vectors with HVR7 from the low seroprevalence adenovirus, Ad26 swapped into the Ad5 vector, we used the following strategy.

1. Using the Ad26 cosmid, PCR amplify a fragment (fragment 1; 131 bp) using the oligonucleotides 9 and 12 to generate the Ad26 HVR7 fragment with Ad5 overlaps at either end.
2. Using the Ad5 cosmid (or hexon shuttle plasmid), PCR amplify an Ad5 hexon fragment (fragment 2; 575 bp) from the *Nde* I site until HVR7 using oligonucleotides 7 and 10.
3. Using the Ad5 cosmid (or hexon shuttle plasmid), PCR amplify an Ad5 hexon fragment (fragment 3; 1,391 bp) from HVR5 until the *Bam* HI site using oligonucleotides 8 and 9.
4. By PCR, fuse fragment 1 (131 bp) with fragment 2 (575 bp) using oligonucleotides 7 and 12, resulting in a fragment (fragment 4) of 686 bp.
5. By PCR, fuse fragment 4 (686 bp) and fragment 3 (1,391 bp) by PCR using oligonucleotides 7 and 8 resulting in a hexon fragment of 2,059 bp containing the HVR7 of Ad26 (*see* **Note 4**).
6. Using the restriction endonucleases *Bam* HI and *Nde* I, subclone the generated fragment back into the hexon shuttle plasmid.
7. Sequence the insert to confirm presence of the correct modification, and to check no other mutations have been introduced by the PCR procedures.
8. For homologous recombination, linearize the hexon shuttle vector using *AsiSI*, and rescue the full length genome using BJ5183 electroporation competent cells, as described previously (*see* **Note 3**).
9. Rescue recombinant adenoviral vectors using permissive cells (e.g. 293 cells) as described previously (*see* **Note 3**).

3.1.5 Generating Ad5 Containing HVR 5 from Ad26

To generate Ad5 vectors containing the HVR5 from Ad26, we used a similar strategy to above.

1. Using an Ad26 genome as a template, generate Ad26 HVR5 fragment (fragment 1; 69 bp) using oligonucleotides 13 and 16
2. Using the Ad5 cosmid (or hexon shuttle plasmid), amplify an Ad5 hexon fragment (fragment 2; 116 bp) from the *Nde* I site until HVR5 using oligonucleotides 7 and 14.
3. Using the Ad5 cosmid (or hexon shuttle plasmid), amplify an Ad5 hexon fragment (fragment 3; 1,898 bp) from HVR5 until the *Bam*H I site using oligonucleotides 15 and 8.
4. By PCR, fuse together fragments 1 (69 bp) and fragment 2 (116 bp) using primers 7 and 14 resulting in a fragment of 168 bp (fragment 4).
5. By PCR, fuse fragment 4 (168 bp) and fragment 3 (1,898 bp) using oligos 7 and 8 resulting in a hexon fragment of 2,050 bp containing the HVR5 of Ad26 (*see* **Note 4**).
6. Using the restriction endonucleases *Bam* HI and *Nde* I, subclone the generated fragment back into the hexon shuttle plasmid.
7. Sequence the insert to confirm presence of the correct modification, and to check no other mutations have been introduced by the PCR procedures.
8. For homologous recombination, linearize the hexon shuttle vector using *Asi*SI, and rescue the full length genome using BJ5183 electroporation competent cells, as described previously (*see* **Note 3**).
9. Rescue recombinant adenoviral vectors using permissive cells (e.g. 293 cells) as described previously (*see* **Note 3**).

3.2 Adenoviral QC Using Silver Staining and NanoSight™

3.2.1 Silver Staining

1. Denature 5×10^{10} vp (gauged by microBCA assay for total viral protein) of adenovirus in as small a volume as possible (ideally less than 50 μl) in denaturing buffer by heating at 95–100 °C for 10 min.
2. Subject denatured adenovirus to electrophoresis on a 9 % polyacrylamide gel for approximately 2–3 h, until the viral proteins/ladder reach the end of the gel (*see* **Note 5**).
3. Carefully remove the gel from the casing, wash with dH_2O, place into plastic box, and stain a commercial silver stain kit following the manufacturer's instructions.
4. Following destaining to reveal banding pattern, compare banding pattern with positive control to assess for any defects in packaging of modified Ad vectors compared to wild type capsid configuration (*see* Fig. 2).

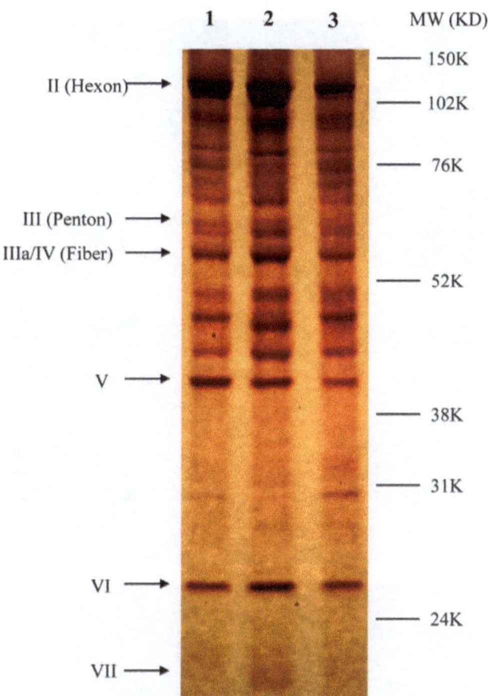

Fig. 2 Silver staining of Adenoviral capsid proteins. Composition of Ad5 and hexon modified adenoviral vectors analyzed by silver stain. 5×10^{10} denatured viral particles were loaded per lane and run on a 9 % SDS-polyacrylamide gel, and stained using Pierce silver staining kit following manufacturer instructions. Lane (1) Ad5-CMVLacZ, Lane (2) Ad5-CMV-LacZ with point mutations in HVR5, Lane (3) Ad5-CMV-LacZ with point mutations in HVR7

3.2.2 NanoSight™ for Assessing Viral QC

We have recently been using the NanoSight™ platform as a tool for assessing viral quality and titers, and find that it accurately reflects total particle count as gauged by viral microBCA protein assay. It also provides additional information on particle monodispersity and potential aggregation, and we have also found it to be a useful tool for studying virus: host protein interactions. The following gives an overview of how we use the machine, however, we recommend full, hands-on training from a NanoSight™ expert, as well as careful reading of the provided manual prior to commencing experiments. Example screen shots and data are shown in Fig. 3.

1. With the computer switched on, open NTA software.
2. Dilute virus prep 1/1,000 in PBS, so the concentration is within the defined detection range (10^6 to 10^{10} particles per mL—occasionally further dilution will be necessary). A minimum total volume of ~1,000 μL is required (hence 1 μL of viral stock solution) (*see* **Note 6**).
3. Undo screws holding the lid in place, clean both glass plates with ethanol and lint free paper, then replace the lid and screw back into place.

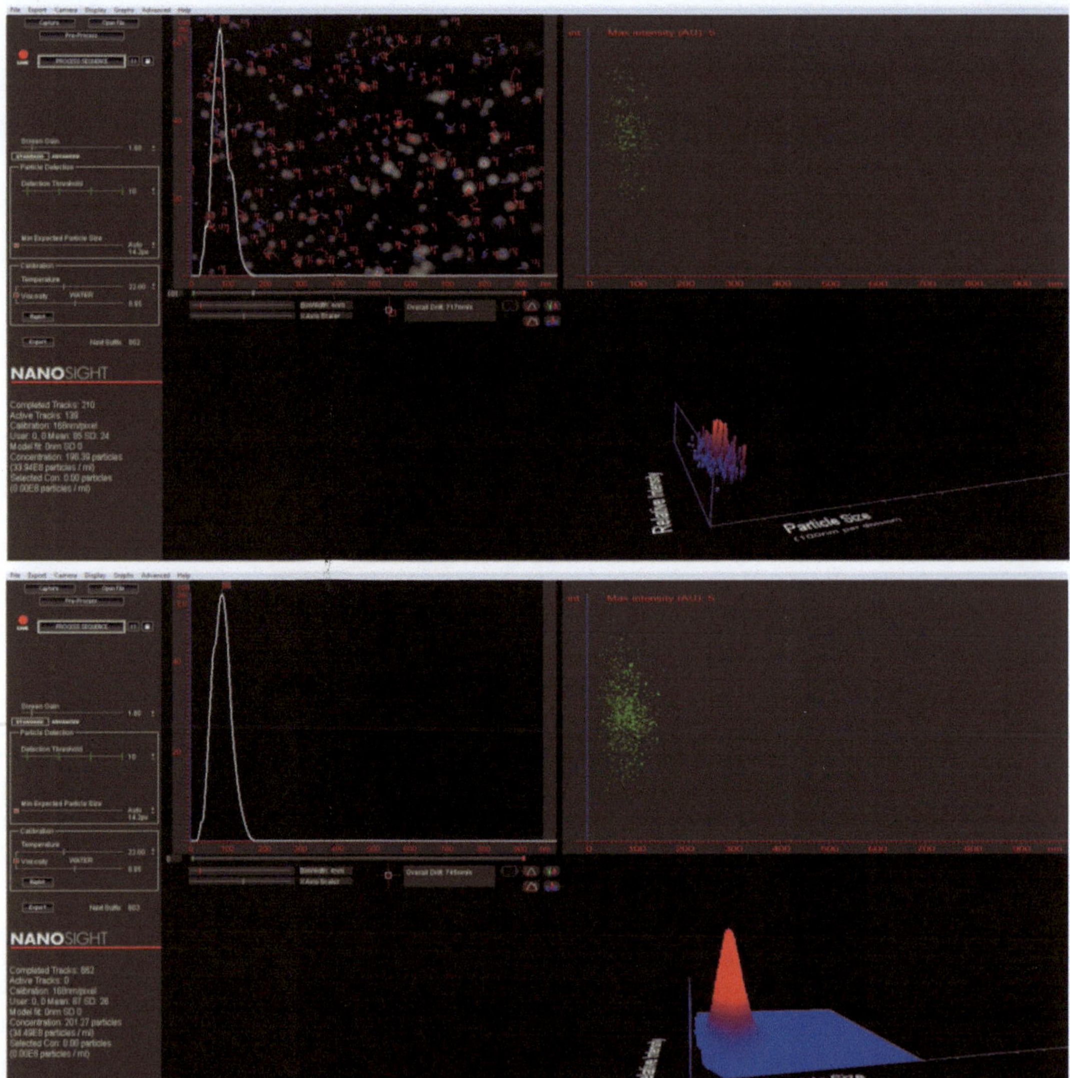

VIRUS	PFU	μBCA	Nanosight™
Ad5 CMVeGFP	2.2×10^{11}	4.6×10^{12}	4.5×10^{12}
Ad5 CMVLacZ	1.25×10^{11}	4.3×10^{12}	3.5×10^{12}
Ad5 CMVLuc	1.88×10^{10}	3.9×10^{12}	3.6×10^{12}

Fig. 3 NanoSight for determining Adenoviral particle count and polydispersity. Screen grabs of NTA particle quantification and QC using NanoSight ™ technology during (*top*) and post (*bottom*) capture of data. *Bottom*—table showing comparison of physical particle counts as gauged by total protein (μBCA) or NanoSight™. Also shown are infectious units as gauged by PFU assay

4. Take up virus solution into syringe, ensure that solution is bubble free, and inject virus into the cell until the glass section is covered (*see* **Note 7**).
5. Place block on microscope, ensure that the laser is on, and open shutter.
6. Look for virus movement which will be noted as small dancing particles in the green beam clearly (*see* **Note 8**). The "sweet spot" for visualizing virus is just to the right of the larger space (*see* **Note 9**).
7. Close shutter, click capture on the NTA software. Adjust the gain/exposure and refocus.
8. Record for 60 s, polydispersity medium and concentration medium (*see* **Note 10**).
9. After movie has recorded, the gain can be adjusted again if needs be. Autotracking option will focus on the center of each particle. Once happy, click "Analyse" (*see* **Note 11**).
10. Record three different parts of the beam and average results (vp/mL).
11. Set report options and save/print using CutePDF (*see* **Note 12**).

4 Notes

1. We normally use gel extraction kits from Promega or Invitrogen, and find both work well, with reasonable recoveries.
2. As a general rule, our laboratory uses 293 cells for viral propagation, however alternative E1 complementing cell lines can be used, for example 911 cells or PERC6 cells.
3. Extensive notes on these protocols, giving full details are found in our previous chapter—please *see* ref. [25].
4. Hypothetically, it is possible to fuse all three fragments directly in one PCR reaction using all three PCR products in approx. equimolar amounts, with the flanking primers. However, we have found it more successful to perform two separate PCR reactions.
5. Our laboratory uses commercially available ladders from Promega, which work well, however we assume that any mainstream supplier should work equally well.
6. We normally make up a 2/3 mL working solution in case (as occasionally happens) the run fails.
7. It aids visualization at this point to have the laser switched on.
8. The speed of movement indicates particle size, not how big they look on screen.

9. The arrow in the diagram below show the approximate position of the "sweet spot" for visualizing particles.

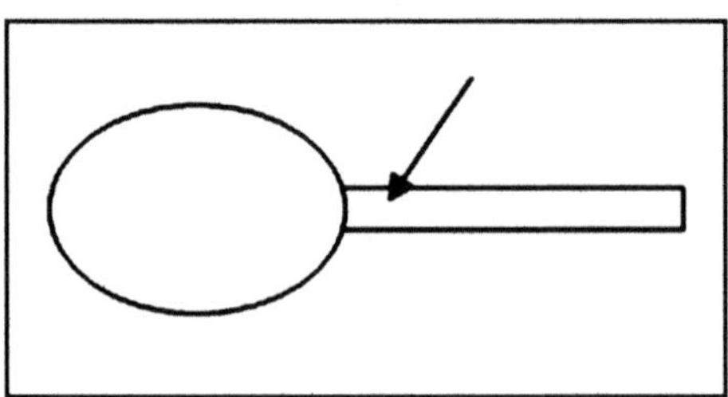

10. The "Batch Capture" option in advanced allows image capturing over a range of timepoints.
11. Values can be excluded by drawing red boxes on the right hand pane. Choose smoother "mountain" graph option, found in middle buttons. User controlled bars can be placed either side of the left hand side graph by using left and right click.
12. Up to ten data sets can be loaded and overlaid onto each graph for comparison.

References

1. Harrach B, Benko M (2007) Phylogenetic analysis of adenovirus sequences. Methods Mol Med 131:299–334
2. Bergelson JM, Cunningham JA, Droguett G, Kurt-Jones EA, Krithivas A, Hong JS et al (1997) Isolation of a common receptor for Coxsackie B viruses and adenoviruses 2 and 5. Science 275:1320–1323
3. Kirby I, Davison E, Beavil AJ, Soh CP, Wickham TJ, Roelvink PW et al (2000) Identification of contact residues and definition of the CAR-binding site of adenovirus type 5 fiber protein. J Virol 74:2804–2813
4. Carlisle RC, Di Y, Cerny AM, Sonnen AF, Sim RB, Green NK et al (2009) Human erythrocytes bind and inactivate type 5 adenovirus by presenting Coxsackie virus-adenovirus receptor and complement receptor 1. Blood 113:1909–1918
5. Seiradake E, Henaff D, Wodrich H, Billet O, Perreau M, Hippert C et al (2009) The cell adhesion molecule "CAR" and sialic acid on human erythrocytes influence adenovirus in vivo biodistribution. PLoS Pathog 5:e1000277
6. Othman M, Labelle A, Mazzetti I, Elbatarny HS, Lillicrap D (2007) Adenovirus-induced thrombocytopenia: the role of von Willebrand factor and P-selectin in mediating accelerated platelet clearance. Blood 109:2832–2839
7. Wickham TJ, Mathias P, Cheresh DA, Nemerow GR (1993) Integrins alpha v beta 3 and alpha v beta 5 promote adenovirus internalization but not virus attachment. Cell 73:309–319
8. Di Paolo NC, Miao EA, Iwakura Y, Murali-Krishna K, Aderem A, Flavell RA et al (2009) Virus binding to a plasma membrane receptor triggers interleukin-1 alpha-mediated proinflammatory macrophage response in vivo. Immunity 31:110–121
9. Bradshaw AC, Coughlan L, Miller AM, Alba R, van Rooijen N, Nicklin SA et al (2012) Biodistribution and inflammatory profiles of novel penton and hexon double-mutant serotype 5 adenoviruses. J Control Release 164:394–402
10. Parker AL, Waddington SN, Nicol CG, Shayakhmetov DM, Buckley SM, Denby L et al (2006) Multiple vitamin K-dependent coagulation zymogens promote adenovirus-mediated gene delivery to hepatocytes. Blood 108:2554–2561
11. Shayakhmetov DM, Gaggar A, Ni S, Li ZY, Lieber A (2005) Adenovirus binding to blood factors results in liver cell infection and hepatotoxicity. J Virol 79:7478–7491
12. Kalyuzhniy O, Di Paolo NC, Silvestry M, Hofherr SE, Barry MA, Stewart PL et al

(2008) Adenovirus serotype 5 hexon is critical for virus infection of hepatocytes in vivo. Proc Natl Acad Sci USA 105:5483–5488
13. Waddington SN, McVey JH, Bhella D, Parker AL, Barker K, Atoda H et al (2008) Adenovirus serotype 5 hexon mediates liver gene transfer. Cell 132:397–409
14. Doronin K, Flatt JW, Di Paolo NC, Khare R, Kalyuzhniy O, Acchione M et al (2012) Coagulation factor X activates innate immunity to human species C adenovirus. Science 338:795–798
15. Alba R, Bradshaw AC, Coughlan L, Denby L, McDonald RA, Waddington SN et al (2010) Biodistribution and retargeting of FX-binding ablated adenovirus serotype 5 vectors. Blood 116:2656–2664
16. Alba R, Bradshaw AC, Parker AL, Bhella D, Waddington SN, Nicklin SA et al (2009) Identification of coagulation factor (F)X binding sites on the adenovirus serotype 5 hexon: effect of mutagenesis on FX interactions and gene transfer. Blood 114:965–971
17. Xu Z, Qiu Q, Tian J, Smith JS, Conenello GM, Morita T et al (2013) Coagulation factor X shields adenovirus type 5 from attack by natural antibodies and complement. Nat Med 19:452–457
18. Parker AL, Waddington SN, Buckley SM, Custers J, Havenga MJ, van Rooijen N et al (2009) Effect of neutralizing sera on factor x-mediated adenovirus serotype 5 gene transfer. J Virol 83:479–483
19. Abbink P, Lemckert AA, Ewald BA, Lynch DM, Denholtz M, Smits S et al (2007) Comparative seroprevalence and immunogenicity of six rare serotype recombinant adenovirus vaccine vectors from subgroups B and D. J Virol 81:4654–4663
20. Roberts DM, Nanda A, Havenga MJ, Abbink P, Lynch DM, Ewald BA et al (2006) Hexon-chimaeric adenovirus serotype 5 vectors circumvent pre-existing anti-vector immunity. Nature 441:239–243
21. Cheng C, Gall JG, Nason M, King CR, Koup RA, Roederer M et al (2010) Differential specificity and immunogenicity of adenovirus type 5 neutralizing antibodies elicited by natural infection or immunization. J Virol 84:630–638
22. Appledorn DM, McBride A, Seregin S, Scott JM, Schuldt N, Kiang A et al (2008) Complex interactions with several arms of the complement system dictate innate and humoral immunity to adenoviral vectors. Gene Ther 15:1606–1617
23. Kiang A, Hartman ZC, Everett RS, Serra D, Jiang H, Frank MM et al (2006) Multiple innate inflammatory responses induced after systemic adenovirus vector delivery depend on a functional complement system. Mol Ther 14:588–598
24. Adams WC, Bond E, Havenga MJ, Holterman L, Goudsmit J, Karlsson Hedestam GB et al (2009) Adenovirus serotype 5 infects human dendritic cells via a coxsackievirus-adenovirus receptor-independent receptor pathway mediated by lactoferrin and DC-SIGN. J Gen Virol 90:1600–1610
25. Alba R, Baker AH, Nicklin SA (2012) Vector systems for prenatal gene therapy: principles of adenovirus design and production. Methods Mol Biol 891:55–84

Chapter 4

Use of Dodecahedron "VLPs" as an Alternative to the Whole Adenovirus

Pascal Fender

Abstract

During human adenovirus type 3 (Ad3) infection, an excess of penton base and fiber proteins are produced. These form dodecahedral particles composed of 12 pentamers of penton base and 12 trimers of fiber protein. Beside this "natural" expression, the adenovirus dodecahedron can be expressed in the heterologous baculovirus system in two forms: a fiber-devoid dodecahedron made only of 12 penton bases (called base-dodecahedron: Bs-Dd) and the fiber-containing dodecahedron (called penton dodecahedron: Pt-Dd). These particles partly mimic the adenoviral cellular entry pathway but are devoid of genetic information making them an unusual tool for basic research or applications. We report here how these particles are expressed and purified, the labeling method for trafficking studies as well as their use in molecular interaction studies. The potential of these particles for biotechnological applications is under evaluation, making their study a "niche" along side traditional adenoviral vectors.

Key words Adenovirus, Dodecahedron, Vectorology, Virus-like particle, Protein transduction, Trafficking, Receptor identification

1 Introduction

Even knowing that Adenoviruses can be produced at high titer (about 10^{12} vp/mL) compared to other viral vectors (10^8 vp/mL in average), this is still far from being comparable with protein yields in overexpression systems. One way to overcome this limitation consists in obtaining "Virus-Like Particles = VLPs" derived from the expression of one or several viral proteins expressed in heterologous systems such as *E. coli*, yeast, *Baculovirus*, mammalian cells or even plants (for review *see* ref. 1). VLPs have been reported for a number of single- or double-stranded RNA and DNA viruses. To the best of our knowledge, all VLPs reproduce the architecture of the virion capsid with the exception of the adenovirus dodecahedron (Ad-Dd). The adenovirus penton base and fiber form a non-covalent complex located at the 12 vertices of the icosahedral capsid and separated by hexon capsomer facets. Indeed,

Miguel Chillón and Assumpció Bosch (eds.), *Adenovirus: Methods and Protocols*, Methods in Molecular Biology, vol. 1089, DOI 10.1007/978-1-62703-679-5_4, © Springer Science+Business Media, LLC 2014

when the penton base and fiber are overexpressed in the *Baculovirus* expression system, a symmetrical dodecameric structure mimicking the overall "collapsed" virion is obtained [2]. A fiberless dodecahedron called Base-Dodecahedron (Bs-Dd) can be also produced [3, 4], indicating that the dodecamerization information is located in the penton base of some but not all adenovirus serotypes (e.g., Ad3 and Ad7). The production and purification of these unconventional particles can be achieved in a few steps to provide a relatively high concentration (3 mg/mL) which corresponds to two to three orders of magnitude more than the adenovirus titer. Both Pt-Dd and Bs-Dd are efficiently internalized by cells, making them good tools for versatile applications in both fundamental research like receptor fishing, trafficking studies [5–7], and in vectorology such as DNA and protein delivery or vaccinology [8–11]. Both, ease of production and safety properties of these genome-devoid particles, in combination with the good structural and functional knowledge, led to increased interest around these VLPs in the past 5 years [7, 12, 13]. In this chapter, state of the art protocols around adenovirus type 3 dodecahedron are described.

2 Materials

2.1 Production: Cell Culture and Baculovirus Infection

1. High Five cell (Invitrogen).
2. Express Five Serum-Free Medium (Invitrogen).
3. Rotative shaker Certomat (Sartorius).
4. Thermostated incubator.
5. Laminar Flow Sterile Hood.

2.2 Purification and Dialysis

1. Ultracentrifuge.
2. SW41 rotor and bucket set (Beckman).
3. Ultraclear 12 mL ultracentrifuge tubes.
4. HBS-G solution: 20 mM HEPES pH 7.4, 150 mM NaCl, 10 % glycerol (*see* **Note 1**).
5. 50 % Sucrose in HBS-G solution.
6. Dialysis Tube with MWCO 100 kDa.

2.3 Dodecahedron Labeling and Cellular Trafficking

1. Protein in amine-free buffer: PBS, HEPES…no TRIS! (*see* **Note 2**).
2. Hoechst 33258 (Sigma).
3. Fluorescent dyes: Alexa Fluor 488 or 555 (Invitrogen) or Cy5 mono-NHS: (GE-Healthcare).
4. Biotin EZ-Sulfo-NHS (Thermo Scientific).
5. Dialysis Tube with MWCO 100 kDa (*see* **Note 3**).

6. Organelles or cytosqueletton detection (Phalloidine Rhodamine: Sigma; Lysotracker: Invitrogen; Organelle-Lights Rab5: Invitrogen).
7. Glass coverslip of 0.17 mm for fixed cells and 0.17 mm Iwaki dishes with glass bottom for live imaging.
8. Thermostated inverted Olympus microscope IX81 or similar.

2.4 Surface Plasmon Resonance

1. BIAcore 3000 (GE-Healthcare).
2. CM4 or CM5 censorships (GE-Healthcare).
3. HBS buffer: 20 mM HEPES, pH 7.4, 150 mM NaCl.
4. EDC-NHS Amine coupling Kit (GE-Healthcare).
5. Streptavidin.
6. Biotinylated Heparin or other lab-made biotinylated ligand.
7. Desmoglein.
8. Regeneration solutions: 0.05 % SDS in water or 20 mM HCl in water.

3 Methods

Adenovirus dodecahedron can be considered an adenoviral mimic. This is particularly true for receptor interaction studies such as penton base/integrins or fiber/desmoglein [6, 7], but nevertheless one must keep in mind that some different features exist. Among them, interaction with HSPGs has never been reported for the whole Ad3, whereas Ad-Dd interacts at high affinity with molecules of this family [5]. In addition, the role played by the hexon in binding coagulation factors has recently been emphasized, but obviously Ad-Dd behaves differently and can only be used as a negative control for studies in this field. The main advantages of the dodecahedron over isolated capsomers (i.e., pentons, fibers) lie both within its multivalency and its spatial constellation [14]. These properties are especially appreciated for surface plasmon resonance studies as the higher molecular weight of VLPs is an advantage in terms of signal and binding avidity can be investigated. In this chapter, purification of Ad-Dd is described, and methods enabling its detection are reported. Applications of these tools from a molecular level to a cellular scale are shown.

3.1 Production: Cell Culture and Baculovirus Infection

1. High Five Cells are cultured in a 27 °C thermostatic incubator or when available (*see* Fig. 1) in a 27 °C room under gentle shaking on the Certomat.
2. Cells are regularly counted using a Neubauer counting cell by adding an equal volume of trypan blue to an aliquot of the cell culture taken under sterile conditions. The cells are left to

Fig. 1 *Baculovirus* facilities at UVHCI (Unit of Virus Host Cell Interactions in Grenoble). Insect cell is cultured under agitation in a 27 °C thermostated room

grow until a concentration of about 10^6 cells per mL is reached. 200 mL of culture can be considered a "standard production."

3. Infect cells at a multiplicity of infection "m.o.i" = 4, by adding the *Baculovirus* aliquot under the sterile hood. After 30 min incubation, put the spinner back under agitation for 60 h.
4. 60 h later, under the hood, dispatch the infected cells into 50 mL sterile plastic tubes. Equilibrate and centrifuge the tubes at 600 × *g* for 5 min at room temperature. Discard the supernatant and store the pellet immediately at −20 °C (*see* **Note 4**).
5. For purification, extemporaneously thaw the pellet by adding 500 μL of hypotonic buffer supplemented with protease cocktail inhibitor and pool all the pellets together. Perform three freezing-thaw cycles (−20 °C, +37 °C) and vortex at each thawing step. Transfer to a 1.5 mL Eppendorf tube, equilibrate, and centrifuge at maximum speed (13,000 × *g*) for 5 min at 4 °C. The cell lysate is then set aside for the next step, whereas pellets of broken cells and debris are discarded in the biological trash.

3.2 Purification and Dialysis

1. Prepare 500 mL of HBS-G and 200 mL of a 50 % Sucrose in HBS-G. A range of 50 mL solutions containing 15–40 % sucrose with an interval step of 5 % is then prepared (*see* **Note 1**).
2. In order to prepare the sucrose gradient, 1.8 mL of the 40 % sucrose solution is added to the bottom of two ultraclear centrifuge tubes. Cushions are then created by gentle addition of 1.8 mL of the following solutions, starting by the 35 % and

ending with the 15 % sucrose solution. The separation between each sucrose step should be easily visible to the naked eye due to the refringence difference. Cell lysate containing the dodecahedron (cf: **step 5** of Subheading 3.1) can then be loaded onto the top of the gradient.

3. Insert both the sample and the balance tubes in previously refrigerated SW41buckets, then equilibrate them accurately with HBS-G buffer. These buckets are then placed in the SW41 rotor together with empty buckets at the remaining positions. Centrifugation is then run for 18 h at 4 °C at 280,000 × *g*.
4. At the end of the run, the ultraclear tube is recovered from the bucket. Fractionation is then done from top to bottom by taking 850 μL samples from the top of the upper solution until the bottom is reached (around 14 fractions as shown in Fig. 2) (*see* **Note 5**).
5. Dodecahedron-containing fractions (usually 11–15) are then pooled together and dialyzed against HBS buffer.

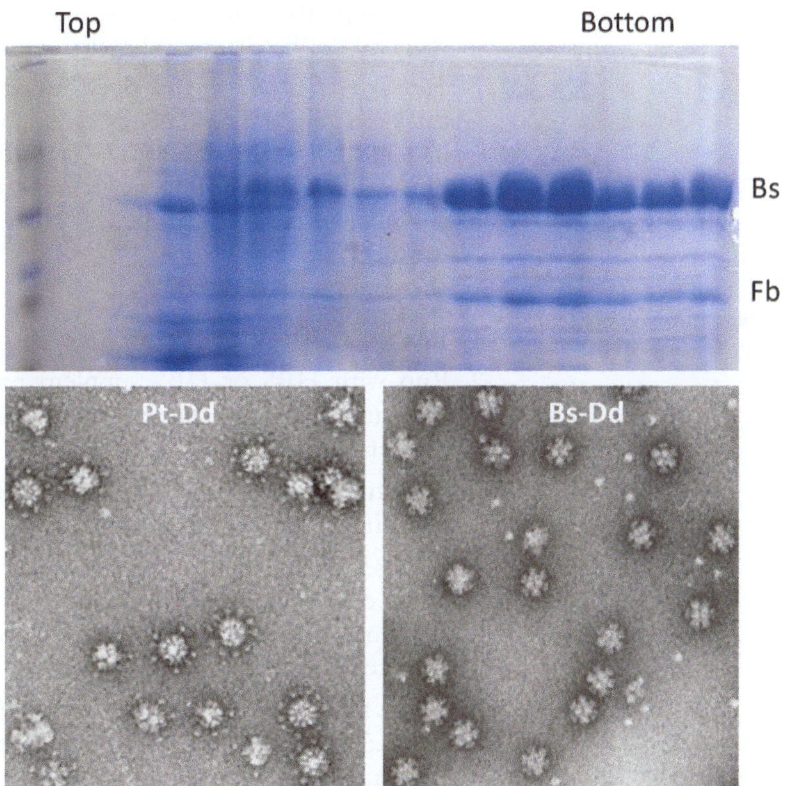

Fig. 2 Dodecahedron purification. *Upper panel*: SDS-PAGE analysis of a portion of each fraction, from top to bottom gradient, after fractionation (*Bs* Base, *Fb* Fiber) Dd is found in bottom fraction. *Lower panels*: Quality control by electron microscopy negative staining of Pt-Dd (*Left*) and Bs-Dd (*right*)

6. DEAE purification can then be performed to eliminate contaminants that are often not visible by SDS-PAGE (i.e., nucleic acids).
7. Quality control can be performed by negative staining electron transmission microscopy (*see* Fig. 2).

3.3 Dodecahedron Labeling and Cellular Trafficking

1. Use dodecahedron at 1 mg/mL in HBS buffer (*see* **Note 2**).
2. Dissolve the dyes in DMSO to get a working concentration of 10 mg/mL.
3. Add 10 μL of dye to 1 mL of dodecahedron solution, mix immediately, and store in the dark at room temperature for 2 h (*see* **Note 6**).
4. Dialyze three times against 500 mL HBS buffer using a high molecular weight cut-off MWCO 100 kDa (*see* **Note 3**).
5. *For fixed cells*: Incubate the coverslip with 1 μg of labeled Dd in 50 μL of medium without serum for the desired period of time. Fix the cells with PFA 2 % for 20 min at 37 °C, then permeabilize for 3 min with PBS—0.1 % triton ×100 (*see* **Notes 7** and **8**).
6. Coverslips are mounted onto the glass slide using a 50 % glycerol/50 % PBS drops and sealed with nail polish.
7. *For live-cell experiments*: Seed Iwaki dishes (0.17 mm glass thick) to get 60 % confluency. Add labeled dodecahedron (2 μg in 200 μL) for 15 min at 37 °C. Remove the sample and add pre-warmed medium. Observation is done in the thermostated chamber of the microscope (*see* **Notes 7** and **9**).

3.4 Surface Plasmon Resonance

For the general procedure:

1. Insert CM5 or CM4 (higher sensitivity) Sensor Chip.
2. Run with HBS or HBS supplemented with 2 mM $CaCl_2$ if calcium is required for the interaction (e.g., cadherins, integrins, etc.).
3. Activate two flow cells with EDC-NHS according to manufacturer instructions. 10 minutes contact is recommended (e.g., 50 μL injection at flowrate 5 μL/min).
4. Short cut the control Flowcell by working only on the “assay” reference Flowcell (e.g., stop Flowcell 1, work only on Flowcell 2).
5. Inject ligands at 10 μg/mL diluted in 10 mM acetate buffer pH 4.5 at 5 μL/min for 10 min. Reinject at the appropriate concentration if immobilization has not reached the expected RU (e.g., 2,000–5,000 RU for ligand with a MW of 60 kDa, 200–500 RU for ligand with 6 kDa MW).
6. Block the chip with a 10-min injection of 1 M Ethanolamine solution on both flow cells.

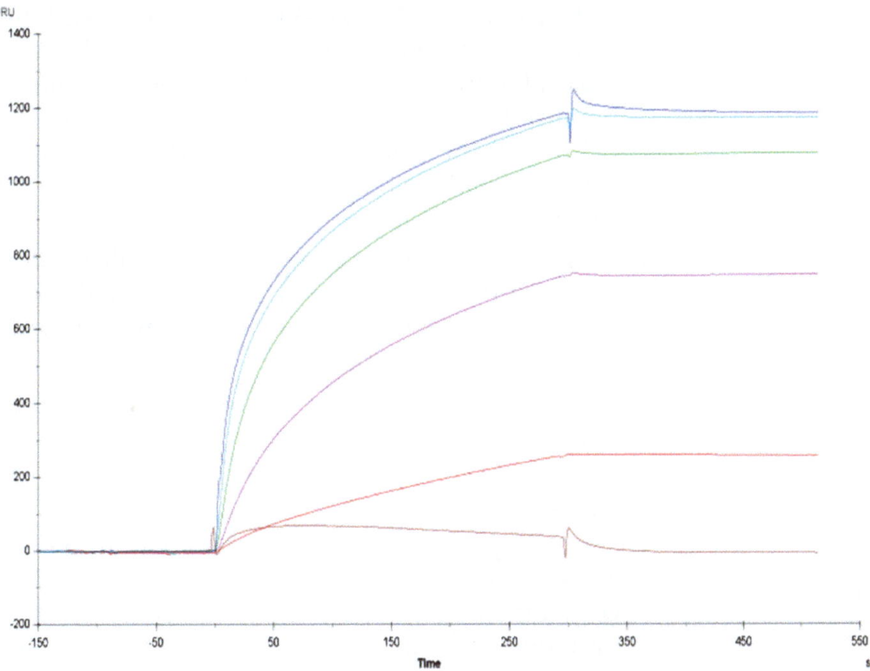

Fig. 3 Surface plasmon resonance experiments of dodecahedron binding to immobilized desmoglein-2 (DSG-2). A range of dodecahedron with fiber (3–100 μg/mL) was injected showing dose–response curves (*red*, *purple*, *green*, *light* and *dark blue*), whereas the fiber-devoid dodecahedron did not show any signal, even at the maximum concentration (*brown curve*). Adapted from Wang and collaborators [7]

7. Inject the analyte on both flow cells from the lowest to highest concentration for 5 min, followed by a 2.5 min dissociation time as shown in Fig. 3. Monitor the result using automatic subtraction of background from the reference cell (*see* **Note 10**).
8. Analyze data by using BIAeval software if applicable (*see* **Note 11**).

4 Notes

1. "Tip" for gradient solution preparation: Add 15 mL of the 50 % sucrose solution to 35 mL of HBS-G to get the 15 % solution, 20 mL of the 50–30 mL HBS-G to get the 20 %, etc. These solutions can be stored at 4 °C for months. Use cool solutions to prepare the gradient to facilitate making clear visible interfaces between each 5 % step.
2. Free amines, even traces are prohibited because chemical reactions target the free amine radical of the lysine lateral chain.
3. "Tip" for dyes with visible emission wavelength: the presence of free dye molecules in the dialysis buffer can be seen by illumination under a UV table. A way to speed up dialysis is by

adding a tube (MWCO 6–8 kDa) with 1 mg/mL BSA together with the tube-containing labeled dodecahedron. The free dye molecules escaping from the Dd tube will rapidly be recaptured and fixed by BSA hence maintaining a low concentration of dye in the buffer and then displacing the equilibrium between inside and outside the tube. This produces a dye-labeled BSA stock.

4. Be gentle when recovering Dd from cell. Centrifuge at low speed (600×*g*) to avoid cell lysis that would result in protein release in the supernatant. Only 2–3 mL can be loaded onto the gradient avoiding the use of supernatant as a sample source.
5. Aliquots of 15 μL of each fraction can be taken to check the presence of the protein of interest (penton base and fiber) by standard SDS-PAGE procedure and Coomassie staining (*see* Fig. 2).
6. The Ad3 penton base monomer weighs 60,000 Da, which is 100× more than the average dye weight (600 Da). The final concentration of the dye is ten times less than the dodecahedron one, meaning ten dyes per monomer. Twenty-five out of the 544 amino acids of the Ad3 penton base are lysines with free amines in the lateral chain. The maximum expected yield is thus (10/25 = 40 %) of labeled lysines per monomer. So finally, the dodecahedron can carry a theoretical maximum of 600 dyes per particle (10 × 5 × 12), giving a good compromise between brightness and non-neutralizing activity of the dye.
7. "Tip" to limit toxicity: Observation can be done using medium with or without serum, PBS is not recommended. When possible, use the highest possible wavelength (i.e., infrared), which results in lower phototoxicity. NB: No toxicity has been observed for Dd observation under "normal" acquisition rates: 3 frames/min for 1 h or 1 min at maximum speed. Cell retraction seen with dodecahedron harboring fiber is a cellular response to fiber signaling as proved by fixed cell controls [14].
8. Optional: Before mounting, perform other labeling using primary antibody (α-actine, α-tubuline, α-clathrin, etc.) and an appropriate secondary labeled antibody. At the end of the experiment the nucleus can be stained by 3 min incubation with Hoechst or propidium iodide as shown in Fig. 4a.
9. Optional: Other live labeling can be performed at the same time (GFP-protein as shown in Fig. 4b, Lysotracker, Hoechst 33342, Organelle-lights, etc.) by following provider instructions. Carry out the acquisition using suitable channels and frame rate (maximum speed for short events or 3 frames per minute for 1 h observation) (*see* **Note 7**).
10. "Tip": We recommend injecting the analyte in the same running buffer to avoid shifts in the curve at the injection start and end points. These shifts result from a short delay in time of

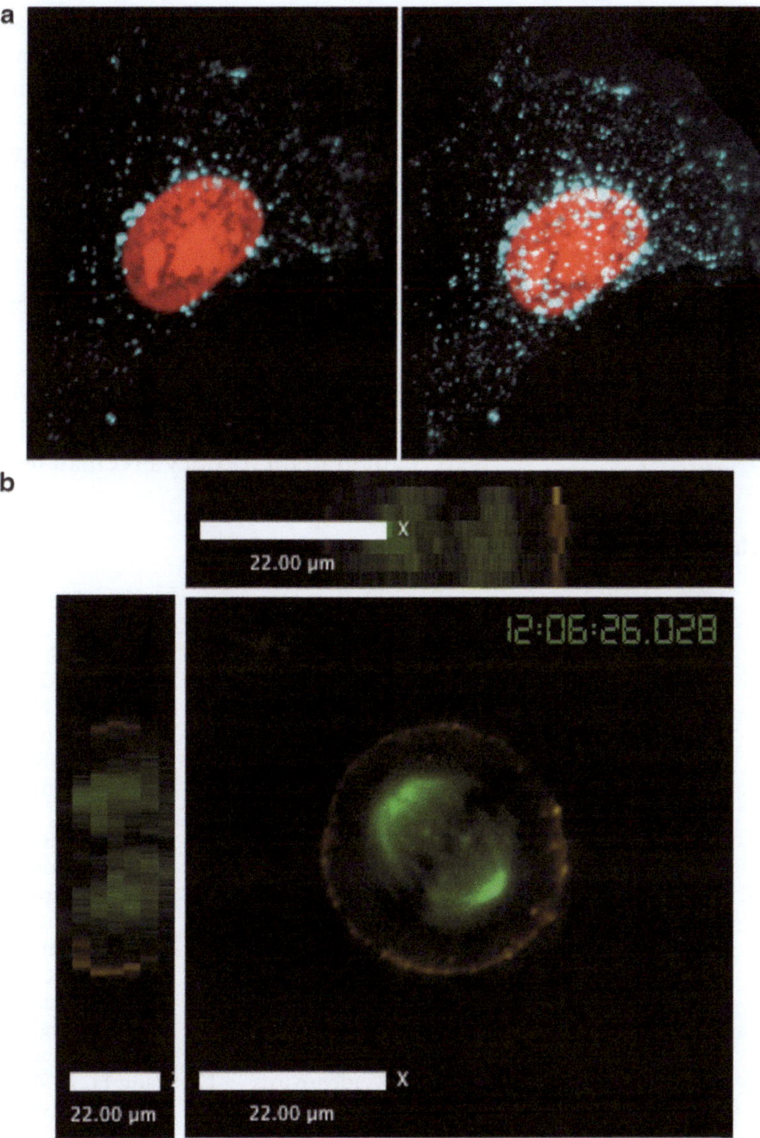

Fig. 4 Labeled dodecahedron in HeLa cells. (**a**) Fixed cell. FITC-labeled Dd was incubated for 1 h. at 37 °C with HeLa cells prior to methanol fixation. Nuclei were counterstained by propidium iodide. Z-series was acquired from bottom to top in both green and red channels with a 0.2 μm step interval. *Left panel* "Bottom slice"; *right panel* "superposition of all Z slices". Note that signal is acquired in gray scale enabling the pseudocoloring of the green FITC in blue. (**b**) Living-cell experiment using HeLa cells expressing GFP-microtubules. Cy3-Dd addition decorates this cell captured in a Z-series. The adjacent panels show XZ and YZ slabs, while medium picture shows one medium XY slice of the Z-series

refractory index between background and reference cells. Ideally, all the samples should be dialyzed in running buffer prior to injection to minimize the shift as shown in Fig. 3, adapted from [12] with permission.

11. Scatchard analysis can be done for complicated analyte/ligand interaction. For this reason, a maximum number of curves must reach a plateau. Maximum binding values at the plateau (in RU) are then plotted along the "*x* axis" while Bound/Free (RU/M "M= Molarity") are plotted along the "*y* axis". The slope "*s*" is then calculated giving a negative number in $(1/M)$. $K_D = -(1/s)$ and is expressed in M.

Acknowledgments

I acknowledge the French "CNRS" for support throughout the dodecahedron story and all the coauthors of publications in this field. I am grateful to Lucy Freeman for English read-through.

References

1. Zeltins A (2012) Construction and characterization of virus-like particles: a review. Mol Biotechnol 53(1):92–107
2. Fender P, Ruigrok RW, Gout E, Buffet S, Chroboczek J (1997) Adenovirus dodecahedron, a new vector for human gene transfer. Nat Biotechnol 15:52–56
3. Schoehn G, Fender P, Chroboczek J, Hewat EA (1996) Adenovirus 3 penton dodecahedron exhibits structural changes of the base on fibre binding. EMBO J 15:6841–6846
4. Fuschiotti P, Schoehn G, Fender P, Fabry CM, Hewat EA, Chroboczek J et al (2006) Structure of the dodecahedral penton particle from human adenovirus type 3. J Mol Biol 356:510–520
5. Vives RR, Lortat-Jacob H, Chroboczek J, Fender P (2004) Heparan sulfate proteoglycan mediates the selective attachment and internalization of serotype 3 human adenovirus dodecahedron. Virology 321:332–340
6. Gout E, Schoehn G, Fenel D, Lortat-Jacob H, Fender P (2010) The adenovirus type 3 dodecahedron's RGD loop comprises an HSPG binding site that influences integrin binding. J Biomed Biotechnol 2010:541939
7. Wang H, Li ZY, Liu Y, Persson J, Beyer I, Moller T et al (2011) Desmoglein 2 is a receptor for adenovirus serotypes 3, 7, 11 and 14. Nat Med 17:96–104
8. Fender P, Schoehn G, Foucaud-Gamen J, Gout E, Garcel A, Drouet E et al (2003) Adenovirus dodecahedron allows large multimeric protein transduction in human cells. J Virol 77: 4960–4964
9. Garcel A, Gout E, Timmins J, Chroboczek J, Fender P (2006) Protein transduction into human cells by adenovirus dodecahedron using WW domains as universal adaptors. J Gene Med 8:524–531
10. Villegas-Mendez A, Garin MI, Pineda-Molina E, Veratti E, Bueren JA, Fender P et al (2010) In vivo delivery of antigens by adenovirus dodecahedron induces cellular and humoral immune responses to elicit antitumor immunity. Mol Ther 18:1046–1053
11. Villegas-Mendez A, Fender P, Garin MI, Rothe R, Liguori L, Marques B et al (2012) Functional characterisation of the WW minimal domain for delivering therapeutic proteins by adenovirus dodecahedron. PLoS One 7:e45416
12. Wang H, Li Z, Yumul R, Lara S, Hemminki A, Fender P et al (2011) Multimerization of adenovirus serotype 3 fiber knob domains is required for efficient binding of virus to desmoglein 2 and subsequent opening of epithelial junctions. J Virol 85:6390–6402
13. Szolajska E, Burmeister WP, Zochowska M, Nerlo B, Andreev I, Schoehn G et al (2012) The structural basis for the integrity of adenovirus ad3 dodecahedron. PLoS One 7:e46075
14. Fender P, Hall K, Schoehn G, Blair GE (2012) Impact of human adenovirus type 3 dodecahedron on host cells and its potential role in viral infection. J Virol 86:5380–5385

Chapter 5

Study of Adenovirus and CAR Axonal Transport in Primary Neurons

Charleine Zussy and Sara Salinas

Abstract

Vectors derived from the canine adenovirus serotype 2 (CAV-2) possess a high neurotropism and efficient retrograde transport that lead to widespread neuronal transduction in the central nervous system (CNS) of various animals. These abilities are due to the engagement of virions to the coxsackievirus and adenovirus receptor at the surface of neurons, which is linked to the endocytic and axonal transport machineries. The trafficking of CAV-2 and the coxsackievirus and adenovirus receptor (CAR) can be visualized ex vivo by incubating primary neurons (e.g., motoneurons and hippocampal neurons) with fluorescently labeled virions or recombinant viral proteins. Using this approach, we could recapitulate the mechanisms responsible for long-range transport of adenovirus in neurons.

Key words Adenovirus, Coxsackievirus and adenovirus receptor, Axonal transport, Fluorescent-ligands, Live-cell imaging

1 Introduction

Adenoviral-mediated gene transfer requires a clear understanding of virus-receptor interaction that underlines cell tropism. In this line, there is now a growing amount of data regarding the intracellular trafficking mechanisms involved during vector entry, mainly in epithelial and nonpolarized cells. Indeed, characterizing at the molecular level the receptor-mediated entry and signaling of viral vectors is crucial prior their clinical use. Adenoviral vectors have been initially used for brain cell transduction in the early 1990s [1]. In the central nervous system (CNS), human adenovirus-derived vectors (HAdV) can infect numerous cell types including neurons, astrocytes, and oligodendrocytes [2]. On the other hand, vectors derived from the nonhuman AdV CAdV-2 (canine adenovirus type 2) have a high neurotropism and efficient axonal retrograde transport in the CNS of rodents, dogs, and primates that allow a widespread neuronal transduction. In this light, CAdV vectors represent promising tools for both understanding and treat neuronal pathologies [2, 3].

Miguel Chillón and Assumpció Bosch (eds.), *Adenovirus: Methods and Protocols*, Methods in Molecular Biology, vol. 1089, DOI 10.1007/978-1-62703-679-5_5, © Springer Science+Business Media, LLC 2014

The coxsackievirus and adenovirus receptor (CAR) is used by many adenovirus, CAdV-2 included [4, 5]. Regarding its expression in the CNS, early studies showed that its expression peaks during development and drops after birth [6]. It has been suggested that CAR could regulate axonal growth through interaction with members of the extracellular matrix [7]. Its role in some AdVs entry is relatively well characterized where it mainly acts as a primary docking factor as its cytoplasmic tail was shown dispensable for infection. Secondary receptors such as integrins will allow virions to recruit the intracellular machinery necessary for their endocytosis. Once in the cell, acidification during endosomal progression will trigger conformational changes in the capsid and subsequent endosome lysis. Members of the motor protein family will then be recruited by cytoplasmic virions to reach the nucleus and inject their genome [8]. CAdV-2, however, does not seem to use secondary receptors. Its cell tropism is absolutely dependent on CAR [5]. We showed that the molecular mechanisms behind CAdV-2 neurotropism and axonal transport were due to the innate ability of CAR in neurons to be endocytosed and axonaly transported towards the cell body [9] (*see* **Note 1**).

Numerous viruses are using molecules found at synaptic terminals such as peripheral nerve endings to enter and spread in the CNS [10]. Visualizing and characterizing their intracellular trafficking can help to better understand their pathogenesis but also key pathways involved in neuronal homeostasis.

2 Materials

2.1 Virus and Protein Production

1. Phosphate buffer saline.
2. Cesium Chloride (CsCl) dissolved in PBS.
3. Ultra clear centrifuge tubes 14 × 95 mm.
4. Needles 21 G × 1½ and Syringes.
5. Slide A Lyser dialysis cassette 10,000 MWO.
6. Glycerol.
7. *Escherichia coli* strain BL21.
8. LB (Luria Broth) supplemented with ampicillin (100 μg/mL).
9. Isopropyl β-D-1-thiogalactopyranoside (IPTG) dissolved in water at 1 M and stored in aliquots at −20 °C.
10. Protino Ni-TED 1000 kit.
11. His-tagged TEV protease.
12. Ni^{2+}-agarose beads.
13. Cy5 and Cy3 mono-reactive dyes.
14. NAP5 column.

2.2 Cell Culture

1. Dulbecco's Modified Eagle's Medium supplemented with 10 % fetal calf serum and antibiotics (100 g/L Streptomycine/10^8 U/L Penicilline; diluted 2,500 times). For DKZeo culture, nonessential amino acids (NEAA) and G418 (final concentration, 0.5 mg/mL) are added.
2. 0.25 % Trypsin.
3. Turbofect reagent.
4. 10× Hanks Balanced Salt Solution diluted in water and adjusted to pH 7.4 with HEPES.
5. Polyornitine dissolved in sterile water at 1.5 mg/mL, filtered, and stored aliquoted at −20 °C.
6. 1 mg/mL Laminin.
7. Glucose.
8. Leibovitz's L-15 medium.
9. Bovine serum albumine.
10. Deoxyribonuclease I (DNAse) resuspended in Leibovitz's L-15 medium at 20 mg/mL, filtered and stored aliquoted at −20 °C.
11. 50× B27 Supplement.
12. L-Glutamine (200 mM—use at 2 mM).
13. 100× Glutamax.
14. Neurobasal medium.
15. Optiprep®.
16. Culture medium for motor neurons and hippocampal neurons: Neurobasal, Glutamax, B27, 0.5 % horse serum, 10 ng/mL CTNF, 200 pg/mL GNDF, and antibiotics.

2.3 Confocal Microscopy

1. Glass-bottom dishes.
2. Microfluidic chambers can be made accordingly to published methods [11] or bougth at Millipore (AXIS™ Axon Isolation Devices) and covalently bound to glass-bottom dishes [11, 12].

3 Methods

3.1 Production and Labeling of CAdV-2 Vectors

1. ΔE1-deleted CAdV-2 vectors are produced and purified following established protocol [13]. Briefly, 12 μg of linearized pCAV-Cre plasmid are transfected with 24 μL of turbofect in 3 wells of a 6-well plate of DKZeo cells cultivated in DMEM (supplemented with FCS, NEAA, and G418) for 5 days.
2. Cells are then scraped and go through three cycles of freeze/thaw. After centrifugation at 5,000 × *g* at RT for 10 min to eliminate debris, supernatant is applied to a 10 cm Ø dish of

DKZeo 60–80 % confluent overnight under agitation. 36 h after, the same procedure is applied to a new 10 cm Ø and subsequently 3× and then 10 × 10 cm Ø before applying supernatants to 30–50 15 cm Ø dishes for the purification.

3. Cracked cells (resuspended in 40 mL of culturing medium) are then loaded on top of a step gradient (six tubes consisting of 2 mL 1.25 g/mL CsCl + 2 mL 1.45 g/mL CsCl) and ultracentrifuged for 1.5 h at 218,000 × *g* at 18 °C. The opaque band at the interface is taken with a syringe (~ 4 mL for the 6 tubes) and further resuspended in 8–9 mL of 1.32 g/mL CsCl (isopycnic gradient). Following 18 h of ultracentrifugation at 218,000 × *g* at 18 °C, the bottom band is collected in 1 mL and dialyzed three times against PBS (with the last dialysis in PBS-10 % glycerol).
4. Multiplicity of infection (physical particles (pp)/mL) is determined by nanodrop. Infectivity (infectious particle/mL) is tested in DKZEo.
5. For labeling, 2×10^{12} pp are diluted in 100 µL of PBS and added to 100 µL of Cy3 mono-reactive dye and 300 µL of PBS 0.1 M HCO_3 for 30 min at room temperature (RT). To remove unconjugated dye, the solution is then dialyzed overnight against PBS and for 3 h against PBS-10 % glycerol (*see* **Note 2**). Infectivity is tested in DKZEo.

3.2 Production and Labeling of CAdV-2 FK

1. *Escherichia coli* strain BL21 is transformed with a CAdV-2 Fiber knob plasmid containing an N-terminal His_6 tag and a TEV protease cleavage site [14]. After cells growth in LB6 amp at 37 °C, protein expression is induced by IPTG (1 mM) for 1 h. The culture is then centrifuged at 2,000 × *g*.
2. The purification is performed with the kit Protino Ni-TED 1000. Briefly, cells are resuspended in lysis buffer and lysed by sonication. The cell lysate is centrifuged at 10,000 × *g* for 30 min at 4 °C, and the supernatant is loaded on a Ni^{2+}-TED (tris–carboxymethyl ethylene diamine) column. His-Tag proteins bound to the resin are then washed and eluted.
3. To separate the His tag from the fiber knob (FK), FK are incubated for 1 h 30 with His-tagged TEV protease at 30 °C. Uncleaved protein and TEV protease are removed by binding to a Ni^{2+}-agarose beads.
4. A Cy5 mono-reactive dye pack is used to label CAdV-2 FK. 200 µg of FK (200 µL) is mixed with 50 µL of the fluorescent dye (one vial resuspended in 500 µL of H_2O) and 25 µL of PBS 0.1 M HCO_3 and incubated for 30–45 min at RT. Labeled proteins are eluted with PBS on a NAP5 column. The final dye/protein ratio (~2.4) is analyzed by spectrophotometer using a NanoDrop ND-100.

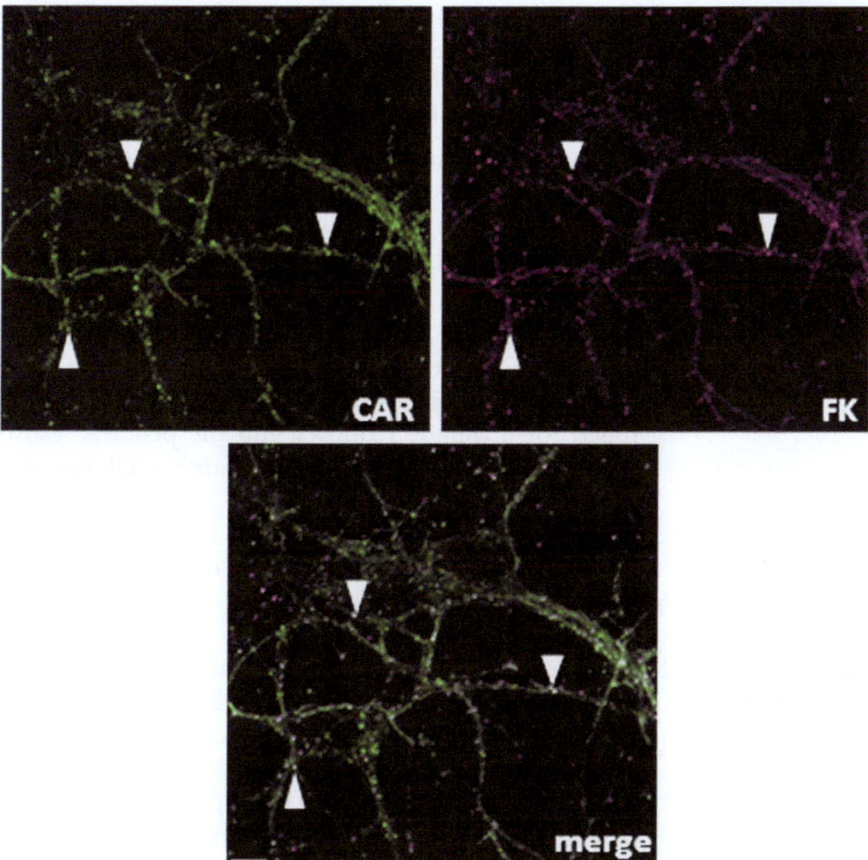

Fig. 1 Hippocampal neurons were incubated with FK for 30 min at 37 °C. After PFA fixation and permeabilization, a specific anti-CAR antibody is used to visualize CAR. *Arrowheads* show colocalization between FK and endogenous CAR. Scale bars: 5 μm

5. To monitor FK interaction with CAR, primary neurons can be incubated with Cy5-FK on ice for 20 min, washed with warm medium and further incubated at 37 °C for 30 min. Figure 1 shows colocalization between FK and CAR in endosomes.

3.3 Culturing Motoneurons and Hippocampal Neurons

1. Primary motoneurons (MNs) (*see* **Note 3**) are isolated either from rat embryonic (E) day 14 or mouse E13 embryos following established protocols [15, 16]. Briefly, spinal cords are removed under a stereomicroscope and trypsined (0.025 %) for 10 min at 37 °C.

2. Cells are resuspended and dissociated in L15 medium + 0.4 % BSA + 0.1 % DNAse and centrifuged on a 4 % BSA cushion at 1,000 × *g*. Cells are resuspended in 1 mL of L15 and selection for MNs is made by centrifugation (for 15 min at 755 × *g* at RT without brake) on a 10.4 % v/v solution of Optiprep® in L-15. The interface is then collected in 1 mL of L15, diluted up to 10 mL, and centrifuged on a BSA cushion. Cells are then

plated in culturing medium on coated glass-bottom dishes (90 min with 15 μg/mL poly-ornithine and overnight with 3 μg/mL of laminin). All imaging experiments are performed using neurons differentiated in vitro for 5–8 days.

3. Primary hippocampal neurons are prepared from E18 mice fetuses. Embryos are taken out and immediately transferred in dissecting media (PBS-Glucose 3 %). Hippocampi (banana-shaped) are microdissected under the stereomicroscope after separation of the two cortical lobes and meninges.
4. Hippocampi are trypsined (0.025 %) for 5 min at 37 °C. Hippocampal cells are then dissociated in 2 mL L15 medium + 0.4 % BSA + 0.1 % DNAse. Cell suspension is centrifuged for 5 min at 1,000 × *g* and the pellet is resuspended in Neurobasal containing B27, L-glutamine, Glutamax, 10 % fetal calf serum, and antibiotics. Cells are then plated on coated glass-bottom dishes (90 min with 15 μg/mL poly-ornithine). All imaging experiments are performed using neurons differentiated in vitro for 7–10 days.

3.4 Live-Cell Imaging and Analyses of CAdV-2 and FK Axonal Transport

1. MNs are incubated with fluorescent ligands (2,000 pp/cells of CAdV2-Cy3 or 1.5 μg/mL of Cy5-FK) for 30–60 min, then washed with warm HBSS (Hanks Balanced Salt Solution)-HEPES pH 7.4. Cells are then imaged by confocal microscopy (e.g., Zeiss LSM 510 or 780 equipped with a 63× 1.4 NA Plan Apochromat oil-immersion objective).
2. After axons are identified morphologically (thinnest process), 100–150 frames are acquired at a rate of 0.4 frame/s (*see* **Note 4**). Images and movies are processed using the LSM 510 or Zen software (Fig. 2a). Kymographs can be generated using MetaMorph (Molecular Devices) by tracing a single line encompassing axons and plotting sequentially every frame of the movie (Fig. 2b).
3. Alternatively, to study axonal transport, microfluidic chambers (MCs, compartmentalized chambers), can be used to isolate fluidically axons from the cell body. In these devices, axons can grow through a parallel array of microchannels connecting two compartments (Fig. 2c, d). MCs are coated with 4 % BSA for 2 h, then 15 μg/mL poly-ornithine for 2 h, and with 50 μg/mL laminin for 1 h 30. All these incubations are performed at 37 °C and three PBS washes are needed between poly-ornithine and laminin. 150,000 hippocampal neurons per MC are then plated in one compartment. To force axonal growth into the microgrooves, BDNF (20 ng/mL final) is added in the axonal compartment after 4 days. 8–10 days after cell plating, fluorescent ligands can be added to the axonal compartment in 150 μL of culturing medium. The cell body compartment

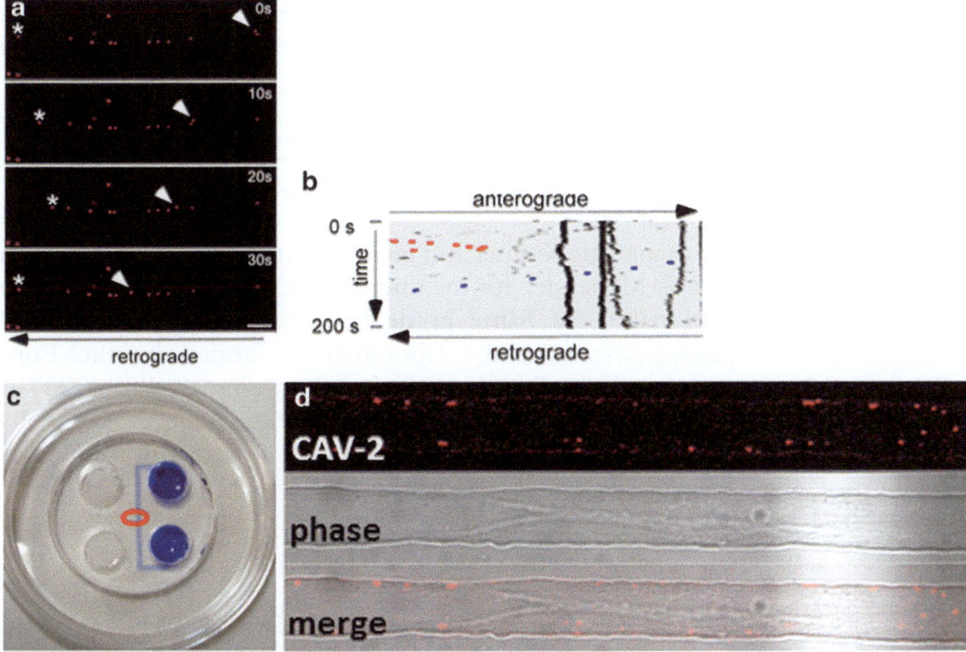

Fig. 2 (**a** and **b**) reproduced from ref. [9]. MNs are incubated with CAdV2-Cy3 for 45 min and imaged by confocal microscopy at 1 frame every 5 s. (**a**) Individual frames are shown. The cell body is located to the *left*. *Arrowheads* show a virion being retrogradely transported, *asterisks* an anterograde virion stopping and changing to a retrograde direction. (**b**) Kymograph of the corresponding movie with a retrograde CAdV-2 highlighted in *blue*. The viral particle labeled by the *asterisk* in (**a**) appears in *red*. (**c**) Picture of a microfluidic chambers. One set of compartment is filled with colorant with a smaller volume than the other compartment. The absence of diffusion of the dye shows the unidirectional flux. *Red oval* highlight the microgrooves. (**d**) Example of transported CAdV2-Cy3 in axons growing in the microgrooves. Scale bars: 5 μm

contains 200 μL of medium to ensure a constant flow towards the axons and avoid diffusion of the probes inside the microgrooves. Axonal transport is visualized using the same parameters than described above.

4 Notes

1. Most of the methods and data generated can be found in ref. [9].
2. All labeled-ligands have to be tested for infectivity/functionality to ensure that the labeling procedure did not affect their structure, which could lead to nonspecific cellular uptake.
3. Neuronal cultures described here are MNs and hippocampal neurons. However, these assays can be applied to virtually any type of neurons that can be cultured in vitro such as for instance dorsal root ganglia neurons, sympathetic neurons, or cortical neurons.

4. Imaging of live transport can be done using various microscopes but we recommend either scanning or spinning-disc confocal with fast rate acquisition.

Acknowledgments

We thank Eric Kremer and members of the EKL group, G. Schiavo, and G. Menendez for the microfluidic chambers. S.S. is an Institut National de la Santé et de la Recherche Médicale (INSERM fellow). Work in the laboratory is funded by the European Community's 7th Framework Programme (FP7/2007–2013; grant 222992—BrainCAV), the French Agence National de la Recherche, E-Rare, the Fondation de France and the Association Française contre les Myopathies.

References

1. Le Gal La Salle G, Robert JJ, Berrard S, Ridoux V, Stratford-Perricaudet LD, Perricaudet M et al (1993) An adenovirus vector for gene transfer into neurons and glia in the brain. Science 259:988–990
2. Kremer EJ (2005) Gene transfer to the central nervous system: current state of the art of the viral vectors. Curr Genomics 6:13–39
3. Bru T, Salinas S, Kremer EJ (2010) An update on canine adenovirus type 2 and its vectors. Viruses 2:2134–2153
4. Coyne CB, Bergelson JM (2005) CAR: a virus receptor within the tight junction. Adv Drug Deliv Rev 57:869–882
5. Soudais C, Boutin S, Hong SS, Chillon M, Danos O, Bergelson JM et al (2000) Canine adenovirus type 2 attachment and internalization: coxsackievirus-adenovirus receptor, alternative receptors, and an RGD-independent pathway. J Virol 74:10639–10649
6. Honda T, Saitoh H, Masuko M, Katagiri-Abe T, Tominaga K, Kozakai I et al (2000) The coxsackievirus-adenovirus receptor protein as a cell adhesion molecule in the developing mouse brain. Mol Brain Res 77:19–28
7. Patzke C, Max KE, Behlke J, Schreiber J, Schmidt H, Dorner AA et al (2010) The coxsackievirus-adenovirus receptor reveals complex homophilic and heterophilic interactions on neural cells. J Neurosci 30: 2897–2910
8. Henaff D, Salinas S, Kremer EJ (2011) An adenovirus traffic update: from receptor engagement to the nuclear pore. Future Microbiol 6:179–192
9. Salinas S, Bilsland LG, Henaff D, Weston AE, Keriel A, Schiavo G et al (2009) CAR-associated vesicular transport of an adenovirus in motor neuron axons. PLoS Pathog 5:e1000442
10. Salinas S, Schiavo G, Kremer EJ (2010) A hitchhiker's guide to the nervous system: the complex journey of viruses and toxins. Nat Rev Microbiol 8:645–655
11. Taylor AM, Rhee SW, Jeon NL (2006) Microfluidic chambers for cell migration and neuroscience research. Methods Mol Biol 321:167–177
12. Taylor AM, Blurton-Jones M, Rhee SW, Cribbs DH, Cotman CW, Jeon NL (2005) A microfluidic culture platform for CNS axonal injury, regeneration and transport. Nat Methods 2:599–605
13. Kremer EJ, Boutin S, Chillon M, Danos O (2000) Canine adenovirus vectors: an alternative for adenovirus-mediated gene transfer. J Virol 74:505–512
14. Schoehn G, El Bakkouri M, Fabry CM, Billet O, Estrozi LF, Le L et al (2008) Three-dimensional structure of canine adenovirus serotype 2 capsid. J Virol 82:3192–3203
15. Deinhardt K, Salinas S, Verastegui C, Watson R, Worth D, Hanrahan S et al (2006) Rab5 and Rab7 control endocytic sorting along the axonal retrograde transport pathway. Neuron 52:293–305
16. Bohnert S, Schiavo G (2005) Tetanus toxin is transported in a novel neuronal compartment characterized by a specialized pH regulation. J Biol Chem 280:42336–42344

Chapter 6

The Use of Chromatin Immunoprecipitation (ChIP) to Study the Binding of Viral Proteins to the Adenovirus Genome In Vivo

Yueting Zheng and Patrick Hearing

Abstract

The encapsidation of adenovirus (Ad) DNA into virus particles depends on *cis*-acting sequences located at the left end of the viral genome. Repeated DNA sequences in the packaging domain contribute to viral DNA encapsidation and several viral proteins bind to these repeats when analyzed using in vitro DNA–protein binding assays. In this chapter, we describe a chromatin immunoprecipitation (ChIP) approach to study the binding of viral proteins to packaging sequences in vivo. This assay permits accurate quantification over a wide range of DNA concentrations. The use of formaldehyde cross-linking to stabilize DNA–protein and protein–protein complexes formed in vivo allows the identification of macromolecular complexes found in living cells.

Key words Chromatin immunoprecipitation, ChIP, DNA–protein complex, Adenovirus, Real-time PCR, qPCR

1 Introduction

The encapsidation of adenovirus (Ad) DNA into virus particles depends on *cis*-acting sequences located at the left end of the genome (Ad5 nucleotides 230–380) [1]. Seven repeated sequences, termed A repeats due to their AT-rich content, are located within this domain that contribute to viral DNA packaging. A repeats A1, A2, A5, and A6 are the most important repeats for packaging activity. A repeats contain a bipartite consensus motif (5′-TTTG N_8 CG-3′). Both the first and the second half-site of the consensus motif, as well as the eight base pair spacing between the half sites, are critical for viral DNA packaging [2, 3]. Two viral proteins, L1 52/55 K and IVa2, have been found to play important roles in Ad packaging and virus assembly, although their exact roles in this process remain unclear [4–7]. The Ad L1-52/55 K protein is found within immature virus particles and this protein forms a physical complex with the Ad IVa2 protein [4].

Miguel Chillón and Assumpció Bosch (eds.), *Adenovirus: Methods and Protocols*, Methods in Molecular Biology, vol. 1089, DOI 10.1007/978-1-62703-679-5_6,

In turn, the Ad IVa2 protein is essential for virus assembly and the formation of empty viral capsids [6]. Ad IVa2 is found in both empty and mature virus particles. The IVa2 protein in vitro binds to packaging A repeats 1 and 2 as well as A repeats 4 and 5 [8]. Furthermore, our recent study suggests that another two Ad proteins, IIIa and L4-22 K, can interact with the packaging sequences in vivo and are required for efficient virus encapsidation [9, 10].

In this Chapter, we describe a chromatin immunoprecipitation (ChIP) approach that was used to study the binding of the Ad IVa2, L1-52/55 K as well as IIIa and L4-22 K proteins to wild type and mutant packaging sequences in vivo using specific antisera directed against these products [11]. The method represents adaptations derived from protocols described previously [12–14]. Viral chromatin was sheared by sonication to an average size of ~200–500 bp and the products of immunoprecipitation were quantified using real-time quantitative PCR (qPCR). This assay permits accurate quantification over a wide range of DNA concentrations. The use of formaldehyde cross-linking to stabilize DNA–protein and protein–protein complexes formed in vivo allows the identification of macromolecular complexes found in living cells.

2 Materials

2.1 Cells and Medium

1. N52.E6 cells (which express Ad E1A and E1B proteins) [15].
2. Dulbecco's Modified Eagle's Medium supplemented with 10 % bovine fetal serum and penicillin/streptomycin.

2.2 Reagents and Buffers

1. Formaldehyde: 37 % formaldehyde.
2. 1 M Glycine.
3. Phosphate-buffered saline (PBS).
4. Sodium dodecyl sulfate (SDS) lysis buffer: 50 mM Tris–HCl, pH 8.0, 10 mM ethylene diamine tetraacetic acid (EDTA), 1 % SDS.
5. Protease inhibitors (working concentrations in SDS lysis buffer, **item 6**): phenylmethylsulfonyl fluoride (PMSF; 1 mM, Sigma Cat. # P7626), aprotinin (1 μg/μl Sigma Cat. # A1153), and pepstatin A (1 μg/μL Sigma Cat. # P4265).
6. IP dilution buffer: 16.7 mM Tris–HCl, pH 8.0, 167 mM NaCl, 1.2 mM EDTA, 0.01 % SDS, 1.1 % Triton X-100.
7. High salt wash buffer: 20 mM Tris–HCl, pH 8.0, 500 mM NaCl, 2 mM EDTA, 0.1 % SDS, 1 % Triton X-100.
8. Low salt wash buffer: 20 mM Tris–HCl, pH 8.0, 150 mM NaCl, 2 mM EDTA, 0.1 % SDS, 1 % Triton X-100.
9. LiCl wash buffer: 10 mM Tris–HCl, pH 8.0, 0.25 M LiCl, 1 mM EDTA, 1 % SDS, 0.5 % Triton X-100, 1 % sodium deoxycholate.

10. TE buffer: 10 mM Tris–HCl, pH 8.0, 1 mM EDTA.
11. Elution buffer: 100 mM $NaHCO_3$, 1 % SDS.
12. Salmon sperm DNA–protein A Agarose (Millipore Cat. # 16-157).
13. Polyclonal or monoclonal antibodies.
14. 20 mg/mL Protease K (NEB Cat. # P8102S).
15. 5 M NaCl.
16. 0.5 M EDTA.
17. 1 M Tris–HCl, pH 6.5.
18. 10 mg/mL Glycogen.
19. Phenol/Chlorofrom.
20. 3 M Sodium Acetate (NaOAc), pH 5.2.
21. DyNAmo™ HS SYBR® Green qPCR Kit (Thermo Cat. # F-410 L).
22. DNase/RNase-Free Distilled Water.

2.3 Equipment and Other Materials

1. Branson Sonifier 450 with 1/8" microtip.
2. DNase/RNase-free barrier tips.
3. Applied Biosystems 7500 Real-Time PCR system (or equivalent quantitative PCR machine).
4. MicroAmp® Optical 96-Well Reaction Plate (Invitrogen Cat. # 4316813).
5. MicroAmp® Optical Adhesive Film (Invitrogen Cat. # 4311971).

3 Methods

3.1 Cultured Cells and Virus Infections

Any cell line that is permissive for Ad infection may be used in these analyses. In our work, we utilized N52.E6 cells which express Ad E1A and E1B proteins [15]. Cells were grown in alpha modification of Eagle medium supplemented with 10 % bovine fetal serum, 2 mM glutamine, penicillin, and streptomycin. Approximately 10^7 cells are found per 100 mm dish. Include one extra dish to be used solely for estimation of cell number. Purified Adenovirus particles were prepared by CsCl equilibrium gradient centrifugation.

1. Cells were infected with 100 virus particles/cell.
2. After 1 h absorbation at 37 °C, remove the virus inoculum, wash the cells twice, and add fresh medium.
3. Incubate infected cells at 37 °C for 18–24 h, although any suitable time point may be used depending on the nature of the analysis.

3.2 Formaldehyde Cross-Linking

1. Aspirate medium from infected cell monolayers and add 5 mL of pre-warmed serum-free medium per 100 mm dish.
2. Add 37 % formaldehyde directly to culture medium to a final concentration of 1 %. Incubate for 10 min at 37 °C.
3. Quench cross-linking by the addition glycine to a final concentration of 125 mM and incubate 5 min at room temperature.

3.3 Cell Lysis

1. Aspirate medium and wash cells twice using ice-cold PBS solution.
2. Add 600 μL of PBS to each plate, scrape the cells from the dish, and transfer into a clean 1.5 mL microfuge tube.
3. Pellet the cells by centrifugation for 5 min at 1,000 × *g* at 4 °C.
4. Resuspend the cell pellet in 600 μl of SDS lysis buffer containing protease inhibitors (1 mM PMSF, 1 μg/mL aprotinin, 1 μg/mL pepstatin A).
5. Incubate the resuspended cells 10 min on ice. [Note: add protease inhibitors to buffer just prior to use. PMSF has a half-life of approximately 30 min in aqueous solution].

3.4 Sonication (See Note 1)

1. To shear the chromatin, merge the microfuge tubes containing cell lysate in an ice-water bath, and sonicate with four rounds of 30 s pulses with 2 min intervals, output control at five, duty cycle at 50 % (Branson Sonifier 450).
2. Clear the cell lysate by centrifuging at 16,000 × *g* for 10 min at 4 °C.
3. Collect supernatant, divide into 500 μL aliquots in microfuge tubes, and store samples at −80 °C.
4. Remove 50 μL of each sonicated sample, this sample is input DNA. It is used to quantify the DNA concentration and as a control in the qPCR:
   (a) Reverse the cross-linking by adding 500 μg/mL proteinase K, incubate at 65 °C for 4 h (or overnight).
   (b) Recover DNA by phenol/chloroform extraction and ethanol precipitation. Wash DNA with 70 % ethanol, air dry.
   (c) Suspend DNA pellet in 50 μL of TE. Read Absorbance at 260 nm of a 1:20 dilution of the sample.
   (d) Run 2 μg of DNA on a 1 % agarose gel in comparison to molecular weight DNA standards to assess chromatin shearing efficiency.

3.5 Normalization of Input Chromatin by qPCR

qPCR gives accurate quantification of target DNA, which allows one to adjust and standardize the input DNA concentration for each ChIP sample. Take 2 μL input DNA prepared in last step (*see*

Subheading 3.4, **step 4**) as an template to measure viral genome enrichment by qPCR. Go to Subheading 3.8. for details of qPCR protocol.

3.6 Immunoprecipitation and Reversal of Cross-Linking

All following manipulations are performed at 4 °C and all wash buffers contain protease inhibitors.

1. Aliquot predetermined volume of chromatin (100 μg total DNA) to 15 mL conical tubes and dilute cell superant tenfold in IP Dilution Buffer containing protease inhibitors (*see* above).
2. Add 80 μL of salmon sperm DNA–protein A agarose slurry to preclear samples. Incubate for 1 h at 4 °C with gentle rotation. (Protein A agarose is supplied as 50 % slurry in ethanol-containing buffer, it should be washed with IP dilution buffer prior to use).
3. Spin samples at 100×*g* for 1 min at 4 °C. Transfer supernatants to new tubes with 15 μL of polyclonal antibody or 2 μg of monoclonal antibody (*see* **Note 2**). Incubate at 4 °C overnight with rotation.
4. On the next day, add 60 μL of pre-washed salmon sperm DNA–protein A agarose and incubate at 4 °C for 1 h with rotation.
5. Pellet the beads at 100×*g* for 1 min.
6. Carefully aspirate supernatant without disturbing beads.
7. Resuspend beads in 1 mL low salt wash buffer and transfer into a clean 1.5 mL microfuge tube. Incubate the samples at 4 °C for 10 min with rotation, followed by brief centrifugation to pellet beads.
8. Carefully aspire supernant.
9. Using this technique, perform the following additional washes of the samples: once with 1 mL of high salt wash buffer, once with 1 mL of LiCl wash buffer, and twice with 1 mL of TE buffer (*see* **Note 3**).
10. Add 150 μL of freshly prepare elution buffer to elute immune complex. Incubate at room temperature for 15 min with rotation.
11. Centrifuge samples at 1,500×*g* for 1 min and transfer supernatant to another clean microfuge tube.
12. Repeat elution and combine the two elution steps supernatant in the same tube (~300 μL in total).
13. Reverse formaldehyde cross-linking by adding 12 μL 5 M NaCl, 6 μL 0.5 M EDTA, 12 μL 1 M Tris–HCl, pH 6.5, and 7.5 μL 20 mg/mL proteinase K and incubate at 65 °C for 4 h (or overnight).

3.7 Purify Immunoprecipitated Chromatin

a. Recover DNA by phenol–chloroform extraction and ethanol precipitation. (Optional: total yield of DNA from ChIP is very low, it might be very hard to locate the DNA pellet after ethanol precipitation. Addition of 20 μg glycogen carrier helps visualize the DNA pellet).

b. Wash pellets with 70 % ethanol and air dry.

c. Resuspend DNA in 50 μL TE.

3.8 Real-Time PCR (qPCR) (See Notes 4 and 5)

In this context, we use DyNAmo™ Hot Start SYBR® Green qPCR Kit (Thermo Scientific) as a tool to quantitatively measure the immunoprecipitated viral genome on an ABI 7500 real-time PCR machine. Any other SYBR green qPCR kits and real-time PCR machine could be used according to manufacturer's instructions. Use PCR-grade tubes, DNase/RNase-free barrier pipet tips, and DNase/RNase-free ultrapure water for all of the following steps. Always include control reactions: negative control with primers but no input DNA and a series of positive controls as a standard curve.

1. Prepare standard curve DNA. Dilute 1.02 μg Ad plasmid DNA pTG3602 with ddH_2O to 500 μL to obtain 10^8copies/2 μl stock (*see* **Note 6**). Then perform tenfold serial dilutions in ddH_2O ranging from 10^8 copies down to 1 copy/2 μL. Use 2 μL of standard DNA in each reaction.
2. Prepare 2× SYBR green master mix. Add 12 μL 50× Rox into every 1 mL SYBR green master mix, mix well by vortexing. This mixture will serve as a 2× master mix in the next step. Rox is a passive reference dye provided along with SYBR green qPCR kit. The concentration of Rox is vary depending on the model of real-time PCR machine, please refer to manufacturer's instruction before use.
3. Prepare qPCR reaction mix in a proper size tube (Table 1). To ensure the precision of qPCR reaction, every sample, including standard DNA and negative control, should be run in three replicates. Calculate total number of reactions needed, and always prepare 5 % extra reactions.

Table 1
Reaction setup for ABI qPCR instruments

Component	Volume	Final concentration
2× SYBR green master mix	10 μL	1×
Forward primer (10 mM)	1 μL	0.5 μM
Reverse primer (10 mM)	1 μL	0.5 μM
ddH_2O	6 μL	
Total volume 18 μL/reaction[a]		

[a]Template DNA is not included

Table 2
Standard cycling protocol for all ABI real-time instruments

Thermal cycler protocol
Stage 1, Reps = 1
Step 1: Hold @ 95.0 °C for 10:00 (MM:SS)
Stage 2, Reps = 40[a]
Step 1: Hold @ 95.0 °C for 00:10 (MM:SS)
Step 2: Hold @ 60.0 °C for 00:40 (MM:SS)
Dissociation protocol[b]
Stage 3, Reps = 1
Step 1: Hold @ 95.0 °C for 00:15 (MM:SS)
Step 2: Hold @ 60.0 °C for 00:30 (MM:SS)
Step 3: Hold @ 95.0 °C for 00:15 (MM:SS)
Settings
Sample Volume: 20 μL 9,600 Emulation
Data Collection: Stage 2, Step 2 (60.0 @ 00:40)

[a]Default cycling program for a ABI qPCR machine is two-step PCR (combined annealing/extension step @ 60.0 °C. Make sure primers are designed suitably for two-step PCR protocol

[b]Dissociation stage is required for SYBR green dye, and instructed by the instrument manufacturer

4. Aliquot 18 μL reaction mix into each well of 96-well plate.
5. Add 2 μL of templates, standard DNA or negative control, mix by pipetting.
6. Seal the 96-well plate with adhesive film.
7. Briefly spin the plate to collect samples at the bottom of wells. Check each well, and ensure no well has abnormal volume of PCR reaction mix.
8. Turn on the real-time PCR machine, and setup up following cycling protocol (Table 2).
9. Save the file before run, then click "Start" to initiate thermal cycling.
10. When the run is finished, the data will be automatically added and saved to the file in **step 9**.
11. Go to tab "Results", you can analyze data under subtab "Amplification Plot", or view "Standard Curve" and "Dissociation" reports.
12. Export and save subtab "Report" as *.csv file, which can be open by Microsoft® Excel.

4 Notes

1. The first important parameter for ChIP is the quality of the cross-linked DNA which needs to be effectively sonicated in order to assure the relevance of the results with specific antibodies and the target DNA segment of interest. Optimal DNA fragment sizes for ChIP are 200–1,000 bp. DNA sonication conditions must be determined empirically for each cell type and sonicator model. Here are some general tips to ensure efficient sonication. First, use a sonicator with a microtip and perform sonication in a 1.5 mL tube. Avoid sample foaming during sonication, since this will decrease the efficiency of chromatin shearing. To avoid foaming, let the microtip of the sonicator reach the bottom of 1.5 mL microfuge tube. To avoid excessive heating of the sample, immerse the sample into an ice-water bath during sonication, and keep at least 2 min intervals between each round of sonication. Also, sonication using a series of short pulses is more efficient than a single long pulse. The numbers of pulse sets should be optimized by prior experiments.
2. The quality of the antibody used for the chromatin immunoprecipitation reactions is one of the most critical parameters of ChIP. Antibodies of high specificity and affinity are required for optimal results. Preimmune serum should be used to verify specificity with polyclonal antibodies and isotype-matched nonspecific monoclonal antibody should be used to verify specificity with monoclonal antibodies.
3. The ChIP wash conditions are fully dependent on the properties of the antibodies chosen. This should be optimized prior to the experiment. Our experience suggests that the wash conditions provided in this protocol are suitable for most antibodies tested. However, for certain antibodies, it might require more a extensive washing regimen during the immunoprecipitations to reduce background and discriminate positive controls from negative controls. For example, five times wash with RIPA buffer (50 mM Tris–HCl, pH 8.0, 750 mM NaCl, 5 mM EDTA, 0.1 % SDS, 1 % Triton X-100, 0.1 % sodium deoxycholic acid) is required for our IVa2 antibody.
4. Another important parameter for qPCR is that primer pairs contain matching annealing temperatures (generally 55–60 °C) and a high degree of sequence specificity (20 bp for Ad DNA, longer for the analysis of genomic DNA). Additionally, the amplicons should be less than 500 bp.
5. A reliable real-time PCR reaction should complete following criteria. First, standard DNA concentrations should cover 7–8 orders of magnitude, such as from 10^8–10^1 copies of Ad DNA/reaction. The correlation coefficient (R^2) value of the standard

curve should be as close to 1 as possible. Second, PCR efficiency for each primer pairs should be as close to 100 % as possible, equivalent to a slope of −3.32. A good reaction should have an efficiency between 90 % and 110 %, which corresponds to a slope of between −3.58 and −3.10. Third, disassociation analysis is recommended when using SYBR green dye as a detector, it is used to determine the specificity of qPCR. For good qPCR with high specificity, every sample, including the standard DNAs but not the negative control, should show one single sharp overlapped peak in the dissociation analysis (also called a melting curve). Finally, to ensure the precision of the qPCR reaction, each sample should be run in three replicates and have a standard deviation of Cq less than 0.167.

6. The plasmid pTG3602 [16], which contains the whole Ad5 genome, is frequently used as a standard DNA in our qPCR experiments. It can be used for any primers that target the Ad genome. However, any other plasmids or purified viral DNA can serve as standard DNA. The copy number can be calculated using a web tool (http://www.thermoscientificbio.com/webtools/copynumber).

References

1. Ostapchuk P, Hearing P (2003) Regulation of adenovirus packaging. Curr Top Microbiol Immunol 272:165–185
2. Schmid SI, Hearing P (1997) Bipartite structure and functional independence of adenovirus type 5 packaging elements. J Virol 71:3375–3384
3. Schmid SI, Hearing P (1998) Cellular components interact with adenovirus type 5 minimal DNA packaging domains. J Virol 72:6339–6347
4. Gustin KE, Lutz P, Imperiale MJ (1996) Interaction of the adenovirus L1 52/55-kilodalton protein with the IVa2 gene product during infection. J Virol 70:6463–6467
5. Hasson TB, Soloway PD, Ornelles DA, Doerfler W, Shenk T (1989) Adenovirus L1 52- and 55-kilodalton proteins are required for assembly of virions. J Virol 63:3612–3621
6. Zhang W, Imperiale MJ (2003) Requirement of the adenovirus IVa2 protein for virus assembly. J Virol 77:3586–3594
7. Zhang W, Low JA, Christensen JB, Imperiale MJ (2001) Role for the adenovirus IVa2 protein in packaging of viral DNA. J Virol 75:10446–10454
8. Zhang W, Imperiale MJ (2000) Interaction of the adenovirus IVa2 protein with viral packaging sequences. J Virol 74:2687–2693
9. Ma HC, Hearing P (2011) Adenovirus structural protein IIIa is involved in the serotype specificity of viral DNA packaging. J Virol 85:7849–7855
10. Wu K, Orozco D, Hearing P (2012) The Adenovirus L4-22K protein is multifunctional and is an integral component of crucial aspects of infection. J Virol 86:10474–10483
11. Ostapchuk P, Yang J, Auffarth E, Hearing P (2005) Functional interaction of the adenovirus IVa2 protein with adenovirus type 5 packaging sequences. J Virol 79:2831–2838
12. Schepers A, Ritzi M, Bousset K, Kremmer E, Yates JL, Harwood J, Diffley JF, Hammerschmidt W (2001) Human origin recognition complex binds to the region of the latent origin of DNA replication of Epstein-Barr virus. EMBO J 20:4588–4602
13. Wells J, Graveel CR, Bartley SM, Madore SJ, Farnham PJ (2002) The identification of E2F1-specific target genes. Proc Natl Acad Sci U S A 99:3890–3895
14. Nelson JD, Denisenko O, Bomsztyk K (2006) Protocol for the fast chromatin immunoprecipitation (ChIP) method. Nat Protoc 1:179–185
15. Schiedner G, Hertel S, Kochanek S (2000) Efficient transformation of primary human amniocytes by E1 functions of Ad5: generation of new cell lines for adenoviral vector production. Hum Gene Ther 11:2105–2116
16. Chartier C, Degryse E, Gantzer M, Dieterle A, Pavirani A, Mehtali M (1996) Efficient generation of recombinant adenovirus vectors by homologous recombination in Escherichia coli. J Virol 70:4805–4810

Chapter 7

DNA Microarray to Analyze Adenovirus–Host Interactions

Stefania Piersanti, Enrico Tagliafico, and Isabella Saggio

Abstract

Defining the molecular toxicity of viral vectors that *are* or *will be* in use for clinical trials is a prerequisite for their safe application in humans. DNA chips allow high-throughput evaluation of the profile of transduced cells and have contributed to underlining specific aspects of vector toxicity both in in vitro and in vivo assets. With gene chips we have been able to identify vector-specific properties, such as the cell cycle alteration induced by vector genomic DNA, along with the activation of specific innate immune pathways that can be ascribed to viral particles. We herein describe a detailed protocol for the use of gene chips to dissect the toxicogenomic signature of human and canine helper-dependent adenoviral vectors. We suggest specific procedures suited for the study of these viral vectors, but we also give indications that can be applied to different experimental contexts. In addition, we discuss the in silico elaboration of gene chip raw data which is a crucial step to extrapolate biological information from gene chip studies.

Key words DNA chip, Transcriptome, Microarray, Gene therapy, Viral vectors, Adenovirus, Toxicogenomics

1 Introduction

For a correct use of adenoviral vectors in the clinic, comprehensive knowledge of their impact on the target host cell, organ, or organism is required. Viral vector induced activation of the immune system has been extensively analyzed, along with vector direct toxicity on tissues following in vivo administration ([1] and references therein). One further approach to deepen the information on adenoviral vector is to study the molecular impact of adenoviral transduction on the host cell transcriptome. In the last years, DNA microarrays (also called DNA chips or gene chips), consisting of a highly ordered matrix of thousands of different DNA sequences with known identities, have become invaluable tools for the global analysis of gene expression, and besides their use in the study of disease states, of development, or of the consequences of gene disruptions and drug treatments ([2] and references therein), they have also been proposed as a means to define the toxicogenomic signature of different gene therapy vectors in

Miguel Chillón and Assumpció Bosch (eds.), *Adenovirus: Methods and Protocols*, Methods in Molecular Biology, vol. 1089,
DOI 10.1007/978-1-62703-679-5_7,

different targets. Global transcriptional analysis has been applied to the study of human adenoviral vectors, of HIV-1-derived vectors, and also of AAV vectors [3–9]. We used gene chips to dissect the response of hepatic cells to helper dependent (HD) and E1-deleted human adenovirus vectors, finding that they induce a response of equal magnitude, in contrast to that observed at the organism level, but with different specific properties [10]. We compared the effect of HD human (HD-HAd) and canine (HD-CAV-2) adenoviral vectors on human brain cells, and assessed that the toxicity-versus-efficacy ratio would suggest that the canine adenoviral vector is more suited for neuron gene transfer than the human one [10]. We also observed that, in brain cells, HIV-1-derived vectors activate a strong interferon response, differently from human and canine HD adenovectors [10]. Di Pasquale et al. assayed AAV-5-based vector transduction in tumor lines and established a correlation between transduction and transcriptional phenotypes, and with this study the PDGF receptor was identified as a functional ligand of AAV-5 [3].

In sum, chip studies have been proven a useful tool for the understanding of the biology of viral vectors and of their relative toxicity, and have also been helpful to make new findings on the biology of wild type counterpart of viral vectors. However, the analysis of the literature also points on one caveat emerging from chip experiments, directly related to the extreme sensitivity of the transcriptome to external insults, that is the interpretation of chip results can be biased by the specific experimental conditions of each data set. Indeed, if we compare independent studies, not only the nature of the vector but also the experimental conditions, the cell type, the timing of the analysis, and the statistical criteria applied, can influence the transcriptome response, and therefore it can be sometimes difficult to drive general conclusions. We believe that one way to better extrapolate the biology from high-throughput studies is to switch from an approach based on the single gene analysis, to one that evaluates the data sets in terms of pathway modulation and applies meta-analysis criteria, which can better meet the needs for correctly understand genomic data. In this chapter then, we detail not only the experimental aspects of transcriptome analysis but also the in silico approaches for data sets interpretation.

2 Materials

2.1 Cell Culture

1. Complete medium: Dulbecco's Modified Eagle's Medium (D-MEM) supplemented with 10 % fetal bovine serum.
2. Trypsin/EDTA solution: 0.05 % trypsin, 0.48 mM EDTA.
3. F-12 Ham nutrient mixture (SIGMA).

4. Glucose-enriched medium: D-MEM with 4.5 g/L glucose (Gibco).
5. Neurobasal medium (Gibco).
6. Gentamicin solution: 50 mg/mL gentamicin sulfate solution (Gibco).
7. B-27 supplement without antioxidants (minus AO, Gibco).
8. Epidermal growth factor (EGF): 20 μg/mL recombinant human EGF (Preprotech), dissolved in water and stored at −20 °C.
9. Fibroblast growth factor (FGF): 20 μg/mL recombinant human FGF (Preprotech) dissolved in 5 mM Tris, pH 7.6 and stored at −20 °C.
10. Alpha-tocopherol: 1 mg/mL alpha-tocopherol (SIGMA) dissolved in ethanol and stored at −20 °C.
11. Alpha-tocopherol-acetate: 1 mg/mL alpha-tocopherol-acetate (SIGMA) dissolved in ethanol and stored at −20 °C.
12. GlutaMAX solution: 200 mM glutaMAX.
13. Forskolin: 10 mM forskolin (SIGMA) dissolved in ethanol.
14. N6,2′-*O*-Dibutyryladenosine 3′,5′-cyclic monophosphate sodium salt (db-cyclic AMP): 1 M db-cyclic AMP (SIGMA) dissolved in water.
15. poly-L-ornithine: 0.01 % poly-L-ornithine solution (SIGMA).
16. Fibronectin: 1 mg/mL human plasma fibronectin (Chemicon, Millipore).
17. T-25 cm^2 flasks with ventilated cap.
18. T-75 cm^2 flasks with ventilate cap.
19. Cell culture dishes: 60 and 100 mm dishes.
20. Accutase solution (SIGMA) stored at −20 °C.
21. Penicillin/streptomycin: 5,000 U/mL penicillin and 5,000 μg/mL streptomycin to be used 1:100.
22. Phosphate-buffered saline (PBS).
23. Expansion medium: 50 % glucose enriched medium, 50 % F-12 Ham, 2 % B-27 minus AO, 20 ng/mL EGF, 20 ng/mL FGF, 10 μg/mL gentamicin, 1 μg/mL alpha-tocopherol, 1 μg/mL alpha-tocopherol-acetate.
24. Dopaminergic differentiation medium: neurobasal medium supplemented with 2 % B-27 minus AO, 10 μg/mL gentamicin, 2 mM glutaMAX, 100 μM db-cyclic AMP, 10 μM forskolin.

2.2 Titration of HD Adenoviruses by Real-Time PCR

1. DNase I: 1 U/μL DNase I amplification grade (Life Technologies).
2. DNase I reaction buffer: 200 mM Tris–HCl, pH 8.4, 20 mM MgCl2, 500 mM KCl, to be used 1:10.

3. Proteinase K: 20 mg/mL proteinase K (Macherey-Nagel) dissolved in proteinase buffer.
4. Proteinase buffer: 100 mM Tris–HCl, 100 mM EDTA, 2.5 % SDS, pH 8
5. SYBR Green Real-Time PCR Master Mix, to be used 1:2 (Life Technologies).
6. HD Human adenovector (pC4HSU and pC4HSUgfp; [11]) 5 μM primers.

   HD-V For 5′ ccaccactacatagcccacagt 3′,

   HD-V Rev 5′ acaaagaatggctgagcaagc 3′.
7. HD canine adenovector (HD-CAV-2 [12]) 5 μM primers:

   GFP For 5′ caacagccacaacgtctatatcatg 3′,

   GFP Rev 5′ atgttgtggcggatcttgaag 3′.
8. Optical 96-well reaction plates (Life Technologies).
9. Optical adhesive films (Life Technologies).
10. Real-Time PCR System 7300 (Life Technologies).

2.3 RNA Extraction, Analysis, and Quantification

1. Qiagen RNeasy mini kit.
2. DNase, RNase free, dissolved in water and stored at −20 °C.
3. Cell Scrapers.
4. Agilent RNA 6000 Nano LabChip kit.

2.4 RNA Retro-Transcription

1. Reverse transcriptase: 200 U/μL SuperScript III Reverse Transcriptase (Life Technologies).
2. PCR grade 10 mM dNTPs (Life Technologies).
3. Random hexamers: prepare a 3 μg/μL solution (Life Technologies).
4. RNase Inhibitor: 40 U/μL RNaseOUT™ (Life Technologies).
5. First-Strand Buffer, to be used 1:5 (Life Technologies).
6. DEPC-treated water (Life Technologies).
7. DTT: prepare a 0.1 M solution.

2.5 Gene Expression Arrays

1. Affymetrix Human Genome U133 Plus 2.0 Array.
2. 3′ IVT Express kit.
3. GeneChip hybridization, wash and stain kit.
4. Affymetrix GeneChip Instrument System.

2.6 Gene Chip Validation by TaqMan Assays

1. TaqMan® Gene Ex Assays (Life Technologies) to be used 1:20.
2. TaqMan Real-Time PCR Master Mix (Life Technologies) to be used 1:2.
3. Optical 96-well reaction plates (Life Technologies).

4. Optical adhesive films (Life Technologies).
5. RNase DNase-free water (Life Technologies).
6. Real-Time PCR System 7300 (Life Technologies).

3 Methods

To apply DNA microarray technology for the transcriptome study in response to adenoviral vector transduction, some issues must be taken into consideration to optimize the reproducibility, reliability, and sensitivity of the analysis. First, in order to obtain reproducible data, cells have to be treated with a vector dose that guarantees homogeneous and significant transduction. In human cells, in vitro, we suggest a dosage of 1,000 adenovector genomes per cell that, in average, ensures at least 50 % of transduction for most of the cell types [10]. A second aspect concerns cell cultures: in order to avoid background transcriptome variability due to the different cell donor and/or passage, we suggest to create a cell bank on which to systematically perform experiments. A further aspect to examine, is the number of replicates needed to obtain reliable results. To obtain reliable estimates of modulation between samples and controls in terms of *p*-value and false discovery rates, assuming that outliers do not occur, three to five gene chips per group are usually adequate. If outliers did occur, the sample should be discarded and repeated. Lastly, another key issue to generate reliable transcriptome data concerns the isolation and processing of DNase-free and high-quality RNA.

3.1 Culturing and Differentiation of Cells

1. The hepatocyte-derived cellular carcinoma cell line Huh7 is cultured in complete medium supplemented with penicillin and streptomycin and incubated at 37 °C, 5 % CO_2. When approaching confluence, Huh7 are passaged with trypsin/EDTA to provide new maintenance cultures on 100 mm tissue dishes and experimental cultures on 60 mm tissue dishes. A 60 mm culture dish is required for each experimental point. A 1:3 split of Huh7 cells provides experimental cultures that reach the confluence after 48–72 h.
2. Human midbrain neuronal progenitor cells (hmNPCs) are cultured in expansion medium as in ref. [13]. hmNPCs are maintained in 5 % CO_2, 3 % O_2 and 37 °C in T-25 cm^2 or T-75 cm^2 flasks coated 3 h with 15 μg/mL poly-L-ornithine and 3 h with 4 μg/mL human fibronectin. hmNPCs are passaged when approaching a confluence of 80–90 %. Cells are detached with accutase, harvested with PBS, centrifuged at 600 × *g*, at 20 °C, 10 min, and plated at the density of 30,000 cells/cm^2.

3. hmNPCs are differentiated by passing them in dopaminergic medium when they are 100 % confluent.

3.2 Extraction of Adenoviral Genomic DNA from CsCl Purified Viral Preparations

1. HD adenoviruses are produced as in Chapter 15. 10^{11}–10^{12} CsCl purified optical viral particles are incubated with 5 U RNase-free DNase I in DNase reaction buffer, 1 h at 37 °C, in order to remove contaminating DNA. DNase is then inactivated by 30 min incubation at 75 °C.
2. Viral capsids are disrupted by incubation with 100 μg proteinase in proteinase buffer in a final volume of 200 μL, 1 h at 37 °C. Proteinase is inactivated by 20 min treatment at 95 °C.

3.3 Quantification of HD Adenoviral Vector Genomes

1. A standard curve is generated from *Pme*I-cut pC4HSU or pC4HSUgfp plasmids. Serial dilutions of the plasmids are prepared ranging from 10^2 to 10^7 copies. Plasmid copies are calculated considering the plasmid size (28,060 bp pC4HSU and 29,859 bp pC4HSUgfp), the equivalency between bp and Da (1 bp = 635 Da), and the number of molecules per mole (Avogadro's number, 6.022×10^{23}). For pC4HSU, the molecular weight is calculated as 28,060 × 635 Da/bp = 17,818,100, therefore the molecules are (μg/17,818,100) × 6.022×10^{23}. The reference curve is obtained by linear regression of the quantification Cq values (*y*-axis) versus the log of molecules number evaluated in each dilution (*x*-axis). PCR efficiency is calculated as $E = 10^{(-1/\text{slope})} - 1$. Optimal amplification efficacy is between 90 and 100 % (3.6 > slope > 3.1) [14, 15].
2. For the reference curve, add plasmid dilutions to a master solution consisting in 2× SYBR Green Real-Time PCR master mix, 250 nM of HD-V- or GFP-pair primers, in 50 μL of total volume, which is loaded into the 96-well plate.
3. For the samples, add 5 μL of undiluted and 1:10, 1:100 dilutions of the extracted vector DNA to the master mix in 50 μL total volume. Samples are amplified in triplicate.
4. The optical 96-well plate is sealed with the optical adhesive film and spun down 1 min at 4 °C. The reaction is performed in 7300 Real-Time PCR system under the following conditions: 50 °C for 2 min and 95 °C for 10 min, followed by 40 cycles of 95 °C for 15 s and 60 °C for 1 min.
5. DNA adenovector copy numbers are calculated by replacing the *y* of reference curve with the Cq values of each sample. The total number of molecules is determined as follows: number of vector copies = quantity × dilution factor × 1,000 μL/5 μL, where the quantity is the number of molecules obtained upon *x* extrapolation from reference curve equation, dilution factor is the inverse of dilution of the extracted vector detected, and

the ratio 1,000 μL/5 μL is the dilution factor to calculate the copy number per milliliter, considering that a volume of 5 μL of template was added to the reaction.

3.4 Infections of Cells with HD Adenoviral Vectors

1. Huh7 are maintained in 100 mm culture dishes. 1×10^6 cells are seeded in 60 mm dishes and after 48 h, cells are detached for counting. Cells are incubated with HD adenovectors at a multiplicity of infection (MOI) of 1,000 vector genomes per cell, 1 h, 30 min at 37 °C in D-MEM supplemented with 5 % FBS, rocking the plate every 20 min. Then, cells are washed twice with PBS, and medium is replaced with fresh complete medium until RNA extraction at the different time points.
2. hmNPCs are seeded in T-25 cm^2 flasks and, when 90–100 % confluent, are incubated 5 days with differentiation medium. Then, cells are incubated with adenoviral vectors at an MOI of 1,000 vector genomes per cell, 2 h in differentiation medium at 37 °C, 5 % CO_2 and 3 % O_2, shaking the cells every 20 min. Afterwards, cells are washed twice with PBS and the RNA is collected at the different time points.

3.5 RNA Extraction for Microarray Hybridization

1. Microarray analysis requires a highly purified RNA generally achieved by using columns after the cell lysis and a digestion of contaminating DNA with DNaseI.
2. The commercial Qiagen RNeasy mini kit for RNA cleanup can be used both on plated cells and after TRIzol extraction and includes the step for on-filter DNase digestion to remove genomic DNA.
3. To effectively purify smaller RNA transcripts such as <200 bp appropriate protocols and kits with efficient enrichment of microRNA should be taken into consideration.
4. At desired time points the medium is removed and 1 mL of Qiagen RNeasy mini kit RPL lysis buffer is added to the cells in T-25 cm^2 flask mixing well.
5. Using the cell scraper, cells are collected with lysis buffer on one side of the flask. Holding the flask inclined, the lysis solution is completely aspired and the lysate is transferred to 2 mL tubes.
6. The lysate is further homogenized by vortexing 1 min. One volume of 70 % ethanol is added to the lysate and vortexed thoroughly. 700 μL of the sample are applied to the column placed in a 2 mL collection tube and centrifuged for 15 s at $\geq 8{,}000 \times g$ at room temperature (RT). The flow-through is discarded. This step is repeated by transferring the rest of sample to the column.
7. Add 350 μL of Qiagen RNeasy mini kit RW1 wash buffer to the column and centrifuge for 15 s at $\geq 8{,}000 \times g$ at RT. The flow-through is discarded.

8. The column is incubated 15 min at RT with 80 μL DNase solution consisting in 10 μL of DNaseI and 70 μL of digestion buffer. Then, 350 μL of RW1 buffer are added to the column and centrifuged 15 s at ≥8,000 × *g* at RT. The flow-through is discarded.
9. Add 500 μL of RPE buffer to the column and centrifuge 15 s at ≥8,000 × *g* at RT. The flow-through is discarded.
10. Add 500 μL of RPE buffer to the column and centrifuge 2 min at ≥8,000 × *g* at RT. The flow-through is discarded.
11. The column is further centrifuged 1 min at RT to eliminate contaminating ethanol and to dry the membrane.
12. To elute, the column is placed in a new 1.5 mL collection tube and 50 μL of RNase-free water are pipetted directly to the membrane. After incubation for 1 min, the column is centrifuged 1 min at ≥8,000 × *g* at RT.
13. RNA concentration is determined spectrophotometrically, where 1 OD at 260 nm corresponds to 44 μg/mL, and the quality of the RNA samples may be analyzed on a standard 1 % agarose gel, 1× TAE gel.
14. RNA samples are stored at −80 °C (*see* **Note 1**).

3.6 RNA Analysis and Quantification

1. Use Disposable RNA chips to determine the concentration and purity/integrity of RNA samples using Agilent 2100 bioanalyzer, as described in the detailed protocol provided by Agilent Technologies (http://www.home.agilent.com/-Agilent 2100 bioanalyzer).
2. The electropherogram of a sample for total RNA (eukaryotic) should resemble the one shown on Fig. 1.

3.7 Chip Hybridization and Scanning

1. cDNA synthesis, biotin-labeled target synthesis, GeneChip arrays hybridization, staining and scanning are performed according to the standard protocol and technical manual of instrumentation supplied by Affymetrix (http://www.affymetrix.com/).
2. Affymetrix Command Console Software is used to perform all technical steps of the procedure and to produce the final data. (*.CEL files).

3.8 Microarray Data Analysis

1. Software packages can be used for data analysis. Probe level data from *.CEL files can be obtained, normalized, and converted to expression values using Partek Genomics Suite (Partek, Inc.), following the RMA algorithm [16] or DChip procedure (invariant set) [17].

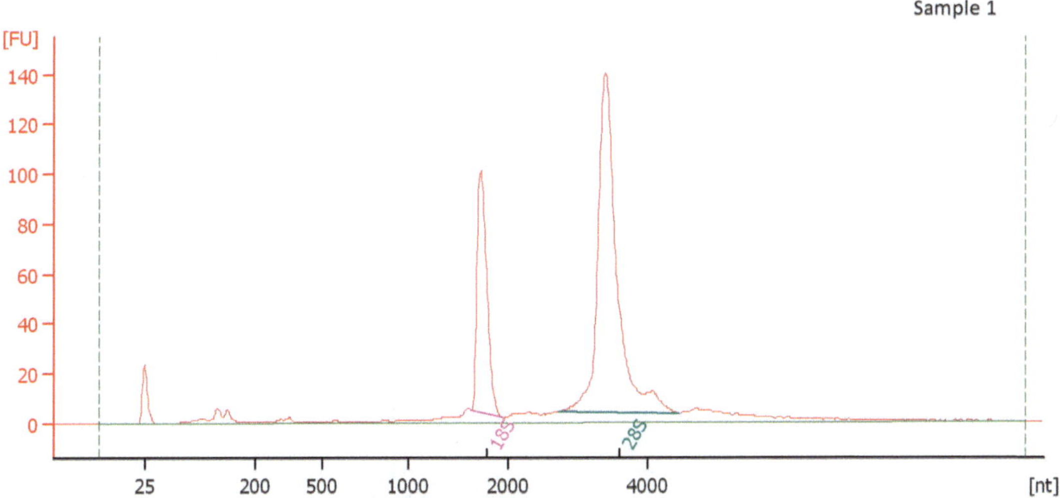

Fig. 1 Agilent Bioanalyzer scan of total RNA from eukaryotic cultured cells isolated using the RNeasy mini Kit from Qiagen. RNA was diluted to a concentration of 100 ng/μL and 1 μL aliquots were analyzed using the RNA 6000 LabChip kit

2. Quality control assessment can be performed using different Bioconductor packages (www.bioconductor.org) such as R-AffyQC Report, R-Affy-PLM, R-RNA Degradation Plot, and Partek's GS (*see* **Note 2**).
3. The DChip-free software can be also used to perform supervised analyses as well as to visualize results.

3.9 Microarray Data Functional Analysis

1. Microarray data analysis measures the changes of gene expression providing a gene list as final output (*see* Fig. 2). However, for a biological interpretation of the experiment, the gene list needs to be elaborated into the format of a functional gene group-based study. Commercial and open source bioinformatic tools are available, that, using public databases (e.g., Gene Ontology and KEGG), permit to systematically dissect large gene lists assembling the transcripts into enriched biological and molecular functional groups. The enrichment analysis is based on the concept that, if a biological process is triggered by a treatment, the related co-functioning genes should have an enriched and higher potential as a relevant group compared to what is represented in the microarray as a whole (gene population background) (*see* Fig. 3). Such over-represented categories represent biological "themes" of a given list. The enrichment can be quantitatively measured by common statistical methods such as Chi-square, Fisher's exact test, binomial probability, and Hypergeometric distribution [18].

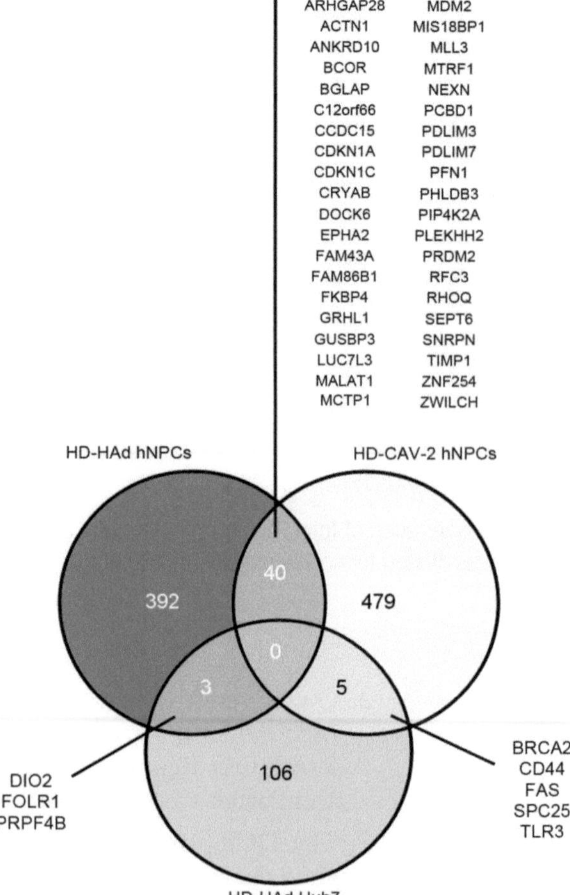

Fig. 2 Venn diagram of genes modulated by HD human and canine adenovectors in human brain and hepatoma cells. The diagram shows the number of common genes modulated by Helper Dependent Human and Canine adenovectors (HD-HAd and HD-CAV-2, respectively) in differentiated hmNPCs and by HD-HAd in the hepatoma line Huh7. Cells are transduced at MOI of 1,000 vector genomes per cell. Transcriptome alteration data were obtained following hybridization of RNA isolated from vector treated- and mock-cells on Affymetrix Gene Chips Arrays. Between HD-HAd and HD-CAV-2, 40 genes are in common in hmNPCs, 3 genes between HD-HAd in hmNPCs and in Huh7, and 5 genes between HD-CAV-2 in hmNPCs and HD-HAd in Huh7. This analysis shows that if only single genes are taken into consideration, the quality of response to the vectors appears strongly dependent on the cell system and experimental conditions

Based on the selected algorithm the enrichment tools are classified as SEA (single enrichment analysis; like EASE and DAVID), GSEA (gene set enrichment analysis; like GoMiner, GFinder, and GProfiler), and MEA (modular enrichment analysis; like GoToolBox) [18, 19].

2. The common pipeline for enrichment analysis consists in uploading a list of pre-selected differentially expressed genes

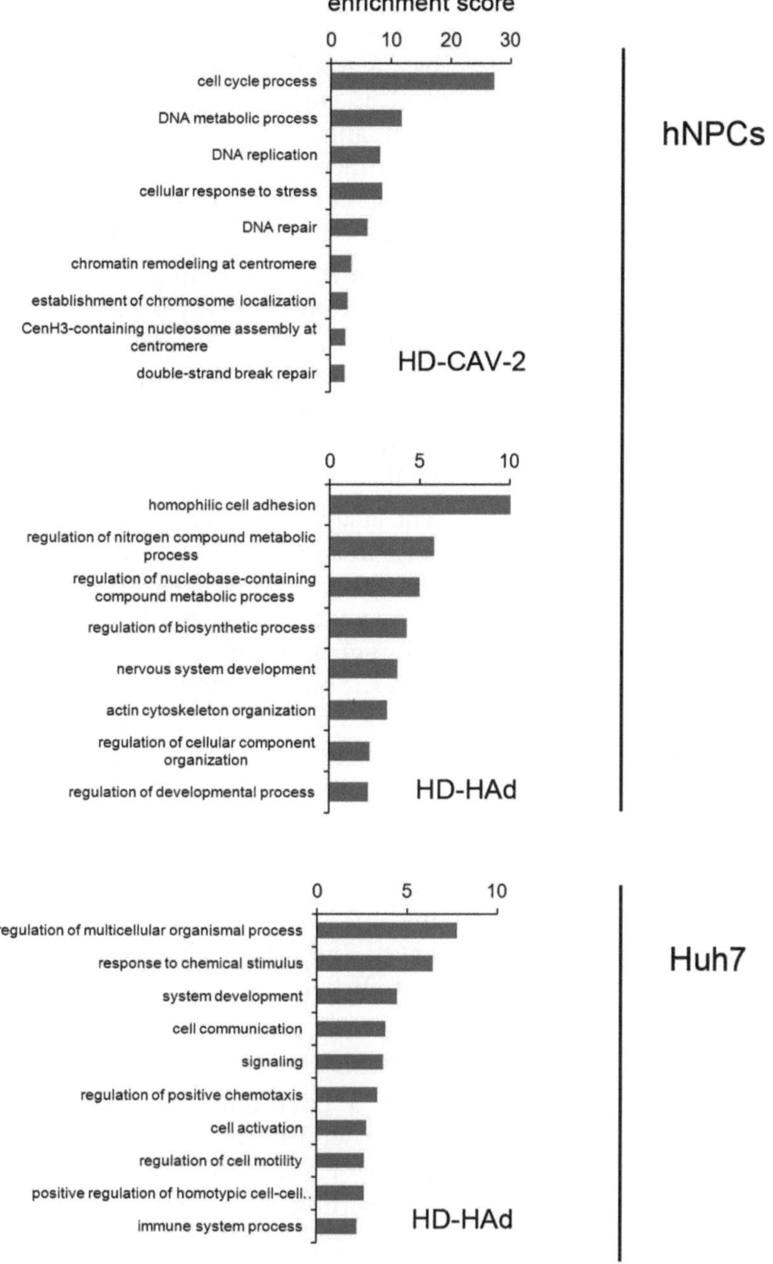

Fig. 3 Enriched biological processes in cells transduced with HD adenoviral vectors. Significantly modulated genes were sorted applying a fold change cut off ≤ -1.5 and $\geq +1.5$ and a *p*-value ≤ 0.05 and were submitted to the Gene Ontology (GO) enrichment classification to find over-represented biological processes regulated by the vector infection. Groups of conceptually similar GO terms were merged so that only the most significant terms are shown applying the threshold Ease score ≥ 2. Biological processes related to cell cycle, DNA metabolism and repair were significantly modified in cells treated with HD-CAV-2, conversely those linked to cell adhesion, system development, cell motility, actin cytoskeleton organization, and immune system were significantly enriched in cells transduced with HD-HAd. This analysis also shows that if gene groups, rather than single transcripts, are taken into consideration, vector specific traits can be recognized in the cell response beyond the background noise of the variations induced by the experimental system

by a T-test with a p-value ≤ 0.05 and a fold change $\geq +1.5$ and ≤ -1.5, and then iteratively testing the enrichment of each term annotation. Alternately, all genes from a microarray experiments are uploaded without any pre-selection of significant genes, in order to allow also minimal transcription changes contributing to the enrichment analysis.

3. Besides the classification of genes based on Gene Ontology, the pathway discovery is another important downstream investigation method, useful to extrapolate biological results. Pathway analysis treats genes in the same pathway as a gene set (*see* **Note 3**). The advantage of analyzing genes as a set is that it can detect coordinate changes that are usually moderate or weak at single gene level.
4. The analysis of differentially expressed genes and related pathways in a given experimental condition can be complemented with insights of putative regulators elements such as transcription factors and microRNAs, based on promoter binding sites and regulative mRNA sequences identified in the co-expressed genes. Some common tools for transcription factor binding site or microRNA prediction and underlying algorithms are Ingenuity Pathway Analysis, oPOSSUM, TFSCAN, and TFSEARCH [19].

3.10 cDNA Synthesis for Microarray Genes Validation by Real-Time PCR

1. Retro-transcription reagents are stored at −20 °C and thawed before use. Add 300–500 ng of purified RNA to the nuclease-free microcentrifuge tube along with the following components: 250 ng of random hexamers, 1 μL 10 mM dNTP mix, and DEPC-treated water up to 13 μL (*see* **Note 4**).
2. Heat mixture at 65 °C for 5 min and incubate on ice for at least 1 min.
3. After a brief centrifugation, add 7 μL/tube of a solution composed of 1 μL 0.1 M DTT, 4 μL 5× First-Strand Buffer, 1 μL of RNase Inhibitor, 1 μL of reverse transcriptase and mix by pipetting.
4. Samples are incubated at 25 °C for 5 min.
5. Next, samples are incubated at 50 °C 50 min.
6. The reaction is inactivated at 70 °C 15 min. The cDNA can now be used as template for amplification by Real-Time PCR.

3.11 Microarray Genes Validation by Real-Time PCR

1. Gene expression changes in the treated samples are measured by relative quantification with respect to the untransduced reference sample. An example of the changes in gene expression induced by three different viral vectors is shown in Fig. 4.
2. Use 1/20 of retro-transcribed cDNA for amplification of the target gene in a solution composed of 12.5 μL of TaqMan®

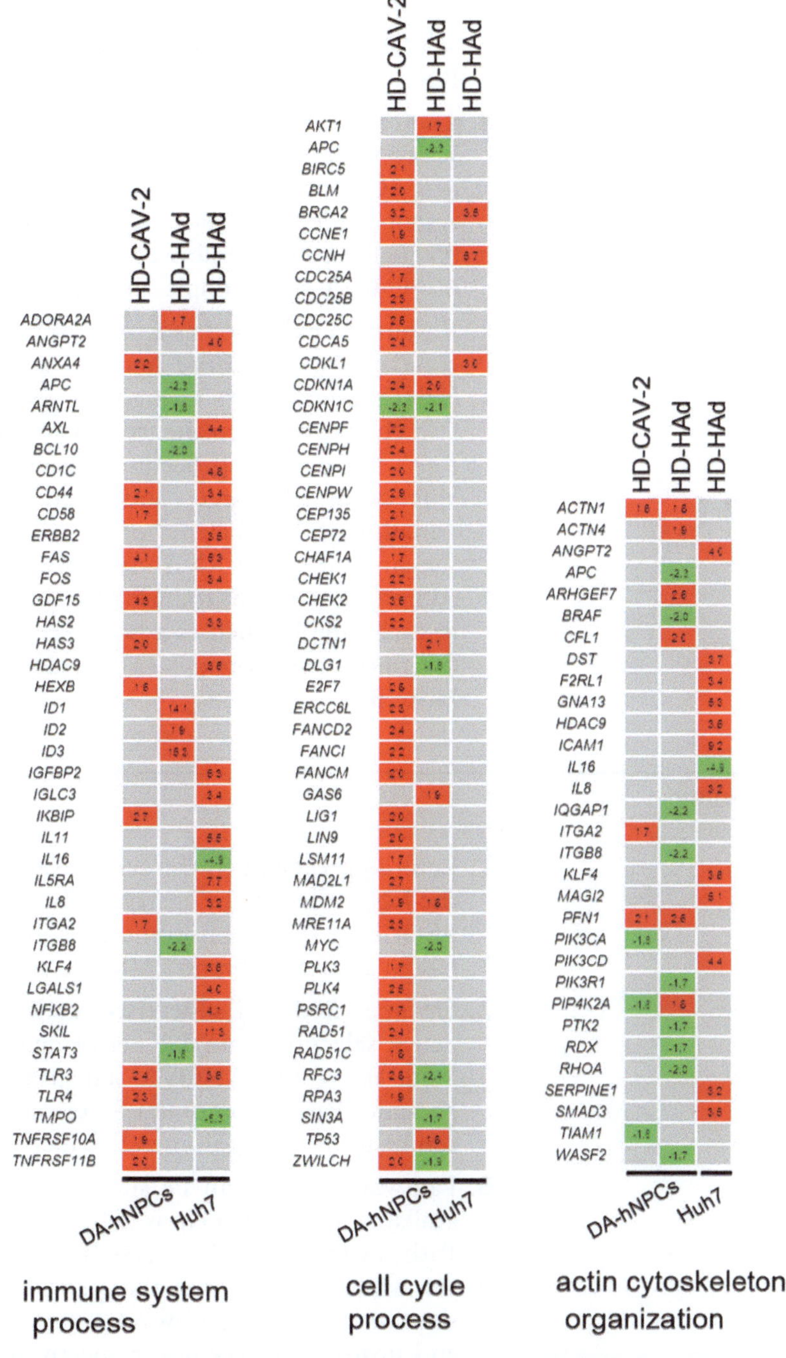

Fig. 4 Heat maps of genes altered in cells transduced with HD adenoviral vectors in selected biological processes. Heat maps represent values and signs of single gene modulations associated to selected biological processes. Genes related to immune system processes and actin cytoskeleton organization were modulated by the three vector systems, but only HD-CAV-2 prevalently induced cell cycle transcripts. *Red*, upregulated genes; *green*, down-regulated genes, *grey*, not modulated genes

Real-Time PCR master mix, 1.25 μL of primers and probe solution of TaqMan® Gene Ex Assays of the selected gene, and RNase DNase free water in a total volume of 25 μL.

3. The reaction mixtures are loaded into optical 96-well plates, sealed with the optical adhesive film, and spun down for 1 min at 4 °C. The reaction is performed in 7300 Real-Time PCR system under the following conditions: 50 °C for 2 min and 95 °C for 10 min, followed by 40 cycles of 95 °C for 15 s and 60 °C for 1 min. Samples are amplified in triplicate.
4. Results are expressed as a treated/reference sample ratio. The comparative procedure applied is the Cq($2\text{-}\Delta\Delta^{Cq}$) method [20], that calculates changes in gene expression as a relative fold change between the treated and the untreated (or calibrator) sample. This procedure assumes that the amplification kinetics of the target and reference gene assays is equal. Variation of the efficiency of the reverse transcriptase as well of the amount of the RNA used for reverse transcription can be addressed by means the use of an endogenous control present in each sample, such as a housekeeping gene like GAPDH or the ribosomal protein L22 encoded by the RPL22 gene (*see* **Note 5**).

4 Notes

1. Before processing to the bioanalyzer, it is preferable to verify the RNA integrity by agarose gel electrophoresis and to quantify it by a spectrophotometer, considering that a pure RNA has an absorbance ratio 260/280 of 1.8–2.0.
2. Unsupervised and supervised analyses (differential expressed genes selection) can be performed using Partek Genomics Suite using the ANOVA statistical implementation of this package. Partek Genomics Suite can be used also to visualize results using the agglomerative hierarchical clustering.
3. Pathway analysis can be supported by commercial and free-available software such as Ingenuity Pathway Analysis (IPA), Pathway studio, PathExpress, and Pathway miner among others, which are platforms that provide genes/proteins relationships as networks, pathways, or explore putative downstream and upstream molecules, respectively, targeted or regulating, the identified pathways [19].
4. The best choice of a priming method (random hexamers, oligodT or gene-specific primers) for reverse transcription should be determined experimentally. Sometimes, especially for long transcripts, a combination of random primers and oligodT is preferable.

5. It is important that the reference gene is expressed in an unchanging fashion regardless the experimental treatment, therefore it is necessary to pre-validate the expression stability of the normalizing gene across the different experimental conditions.

Acknowledgments

This work has been supported by Grants EU FP7 BrainCAV-2 (n. 222992) and FP7 Brainvectors (n. 286071), by the Consorzio Interuniversitario Biotecnologie.

References

1. Nayak S, Herzog RW (2010) Progress and prospects: immune responses to viral vectors. Gene Ther 17:295–304
2. Stoughton RB (2005) Applications of DNA microarrays in biology. Annu Rev Biochem 74:53–82
3. Di Pasquale G, Davidson BL, Stein CS, Martins I, Scudiero D, Monks A et al (2003) Identification of PDGFR as a receptor for AAV-5 transduction. Nat Med 9:1306–1312
4. Haberberger TC, Kupfer K, Murphy JE (2000) Profiling of genes which are differentially expressed in mouse liver in response to adenoviral vectors and delivered genes. Gene Ther 7:903–909
5. Martina Y, Avitabile D, Piersanti S, Cherubini G, Saggio I (2007) Different modulation of cellular transcription by adenovirus 5, DeltaE1/E3 adenovirus and helper-dependent vectors. Virus Res 130:71–84
6. Mitchell R, Chiang CY, Berry C, Bushman F (2003) Global analysis of cellular transcription following infection with an HIV-based vector. Mol Ther 8:674–687
7. Piersanti S, Martina Y, Cherubini G, Avitabile D, Saggio I (2004) Use of DNA microarrays to monitor host response to virus and virus-derived gene therapy vectors. Am J Pharmacogenomics 4:345–356
8. Stilwell JL, McCarty DM, Negishi A, Superfine R, Samulski RJ (2003) Development and characterization of novel empty adenovirus capsids and their impact on cellular gene expression. J Virol 77:12881–12885
9. Zhao H, Granberg F, Elfineh L, Pettersson U, Svensson C (2003) Strategic attack on host cell gene expression during adenovirus infection. J Virol 77:11006–11015
10. Piersanti S, Astrologo L, Licursi V, Costa R, Roncaglia E, Schwarz SC, et al. (2012) Differentiated neuroprogenitor cells incubated with human or canine adenovirus, or lentiviral vectors have distinct transcriptome profiles. Submitted
11. Sandig V, Youil R, Bett AJ, Franlin LL, Oshima M, Maione D et al (2000) Optimization of the helper-dependent adenovirus system for production and potency in vivo. Proc Natl Acad Sci U S A 97:1002–1007
12. Soudais C, Skander N, Kremer EJ (2004) Long-term in vivo transduction of neurons throughout the rat CNS using novel helper-dependent CAV-2 vectors. FASEB J 18:391–393
13. Milosevic J, Adler I, Manaenko A, Schwarz SC, Walkinshaw G, Arend M et al (2009) Non-hypoxic stabilization of hypoxia-inducible factor alpha (HIF-alpha): relevance in neural progenitor/stem cells. Neurotox Res 15:367–380
14. Peirson SN, Butler JN, Foster RG (2003) Experimental validation of novel and conventional approaches to quantitative real-time PCR data analysis. Nucleic Acids Res 31:e73
15. Wong ML, Medrano JF (2005) Real-time PCR for mRNA quantitation. Biotechniques 39:75–85
16. Irizarry RA, Bolstad BM, Collin F, Cope LM, Hobbs B, Speed TP (2003) Summaries of Affymetrix GeneChip probe level data. Nucleic Acids Res 31:e15
17. Castro M, Hurtado-Lorenzo A, Umana P, Smith-Arica JR, Zermansky A, Abordo-Adesida E et al (2001) Regulatable and cell-type specific transgene expression in glial cells: prospects for gene therapy for neurological disorders. Prog Brain Res 132:655–681

18. Demidova AR, Aau MY, Zhuang L, Yu Q (2009) Dual regulation of Cdc25A by Chk1 and p53-ATF3 in DNA replication checkpoint control. J Biol Chem 284: 4132–4139
19. Selvaraj S, Natarajan J (2011) Microarray data analysis and mining tools. Bioinformation 6:95–99
20. Klein D (2002) Quantification using real-time PCR technology: applications and limitations. Trends Mol Med 8:257–260

Chapter 8

Determination of the Transforming Activities of Adenovirus Oncogenes

Thomas Speiseder, Michael Nevels, and Thomas Dobner

Abstract

The last 50 years of molecular biological investigations into human adenoviruses (Ads) have contributed enormously to our understanding of the basic principles of normal and malignant cell growth. Much of this knowledge stems from analyses of the Ad productive infection cycle in permissive host cells. Also, initial observations concerning the transforming potential of human Ads subsequently revealed decisive insights into the molecular mechanisms of the origins of cancer and established Ads as a model system for explaining virus-mediated transformation processes. Today it is well established that cell transformation by human Ads is a multistep process involving several gene products encoded in early transcription units 1A (E1A) and 1B (E1B). Moreover, a large body of evidence now indicates that alternative or additional mechanisms are engaged in Ad-mediated oncogenic transformation involving gene products encoded in early region 4 (E4) as well as epigenetic changes resulting from viral DNA integration. In particular, studies on the transforming potential of several E4 gene products have now revealed new pathways that point to novel general mechanisms of virus-mediated oncogenesis. In this chapter we describe in vitro and in vivo assays to determine the transforming and oncogenic activities of the E1A, E1B, and E4 oncoproteins in primary baby rat kidney cells, human amniotic fluid cells and athymic nude mice.

Key words Adenovirus, Transformation, Oncogenes, E1A, E1B, E4, Tumorigenicity, Baby rat kidney cells, Primary human cells, Human amniotic fluid cells, Nude mice

1 Introduction

Oncogenic transformation of primary cells by human adenoviruses (Ads) is a multistep process that is initiated by the viral E1A gene. Autonomous ectopic expression of E1A gene products is sufficient to induce unscheduled cellular DNA synthesis and proliferation by virtue of their ability to interact with and modulate the function of several growth-regulatory cellular proteins that control transcription and cell cycle progression, including several Rb family members and p300/CBP (reviewed in ref. 1). Consequently, upon E1A expression, primary cells with a limited life span can become immortal. The Ad type 5 (Ad5) E1A transcription unit encodes two major proteins of 289 and 243 amino acids derived from

Miguel Chillón and Assumpció Bosch (eds.), *Adenovirus: Methods and Protocols*, Methods in Molecular Biology, vol. 1089, DOI 10.1007/978-1-62703-679-5_8, © Springer Science+Business Media, LLC 2014

differentially spliced overlapping transcripts, referred to as 13S and 12S RNA, respectively. Whereas the 13S-derived protein (E1A-13S) is required for viral replication, the 12S product (E1A-12S) encodes all functions necessary for immortalization.

In general, E1A-immortalized cells are not completely transformed in that they grow slowly and not to high densities, are anchorage dependent, and usually are not tumorigenic. Instead, full manifestation of the transformed phenotype requires expression of a cooperating second oncogene (reviewed in refs. 2, 3). Such cooperating oncogenes include cellular genes like activated ras. However, within the context of the adenoviral genome, the role of this second oncogenic determinant has been traditionally ascribed to E1B. This transcription unit encodes two major proteins, E1B-55 kDa and E1B-19 kDa, which are structurally unrelated but independently cooperate with E1A proteins in the complete oncogenic transformation of primary cells in culture. This function is, at least in part, a result of their ability to inhibit E1A-induced p53-dependent and p53-independent apoptotic cell death (reviewed in ref. 4). It appears that much of the oncogenic activity of E1B-55 kDa is related to its interaction with the cellular tumor suppressor protein p53, although p53-independent mechanisms are now known to exist. The E1B-19 kDa shares structural and functional similarities with the cellular anti-apoptotic Bcl-2 protein.

Over the past 20 years we and others have shown that, besides E1A and E1B, human Ads contain an additional third oncogenic region, E4 (reviewed in ref. 5). Specifically, the protein products of three open reading frames within this transcription unit—E4orf1, E4orf3, and E4orf6—have been implicated in oncogenic transformation. Work by Javier and colleagues suggests that transformation by E4orf1 differs considerably from the E1-encoded functions and is mediated through a novel mechanism that involves interactions with multiple PDZ-domain-containing proteins [6, 7]. In contrast, the oncogenic activities of E4orf3 and E4orf6 resemble those of the E1B oncoproteins. They can individually cooperate with E1A to completely transform primary rodent cells in vitro. Moreover, they enhance transformation mediated by the E1A and E1B oncogenes. The transforming properties of E4orf6 involve two distinct mechanisms, one of which is linked to degradation of the p53 protein in combination with E1B-55 kDa [8]. The molecular basis of transformation by E4orf3 is not known, but it appears not to be linked to this protein's ability to interact with cellular promyelocytic leukemia bodies (ref. 9 and unpublished results).

The transforming activities of adenoviral E1- and E4-encoded oncogenes can be determined by DNA transfection of primary rodent cells. For this purpose, we routinely use primary kidney epithelial cells prepared from neonatal rats. These cells are commonly referred to as baby rat kidney (BRK) cells. They provide at least two advantages over other cell culture systems, including primary fibroblasts and immortalized cell lines. First, the use of BRK cells results in negligible background transformation. Second, because human neoplasia are predominantly (approx. 90 %) epithelial in origin, the results obtained with BRK (epithelial) cells may be extrapolated to human cancers.

In 2000, Schiedner and coworkers published, that primary human amniotic fluid cells (AFC) are susceptible to Ad5 E1-mediated transformation [11]. Although AFCs are an unspecific mixture of primary human cell types, they can nevertheless be easily obtained after amniocentesis and cultured for at least a couple of weeks in vitro. Therefore, they are quite suitable to study Ad-mediated transformation in a human cell system.

2 Materials

1. Baby rats: a litter (approx 6–12 pups) of 5–7-day-old neonatal outbred Sprague Dawley (SD) rats can be purchased from Harlan SD (Indianapolis, IN).
2. Primary human amniotic fluid cells (AFC) can be easily isolated from the amniotic fluid after amniocentesis for clinical purposes and further cultivated in vitro [11].
3. Nude mice: athymic NMRI (nu/nu) mice can be obtained from Harlan SD.
4. Plasmids: mammalian expression plasmids (e.g., pcDNA3 derivatives) encoding Ad E1A12S/13S, E1B-55 kDa, E1B-19 kDa, E4orf3, E4orf6, and activated *ras* have been described [8, 9] and can be obtained from the authors.
5. Carrier DNA: purified DNA from fish sperm can be obtained from Roche. Resuspend DNA in sterile double-distilled water at 10 mg/mL and shear to an average fragment length of 10,000 bp (evaluated by agarose gel/ethidium bromide electrophoresis) by sonication. Store in aliquots at –20 °C. Dilute 1:10 to make working stocks (1 mg/mL).
6. Lentiviral gene ontology vectors (e.g., LeGO-iV2/Cer2 derivates; Addgene, cat. no. 27334, 27338) encoding Ad E1A12S/13S, E1B-55 kDa, E1B-19 kDa, E4orf3, and E4orf6 can be obtained from the authors.
7. Packaging plasmids pMDLg/pRRE, pRSV-Rev, and pMD2.G can be purchased from Addgene (Addgene; cat. no. 12251, 12253,12259).
8. Producer cell line HEK293T (ATCC; cat.no. CRL-11268TM).
9. Phosphate-buffered saline (PBS) without Ca^{2+} and Mg^{2+}: dissolve 8.0 g NaCl, 0.20 g KCl, 1.60 g Na_2HPO_4 (or 1.78 g $Na_2HPO_4{\cdot}2H_2O$), and 0.24 g KH_2PO_4 in 900 mL of double-distilled water and adjust the pH to 7.2–7.4 with HCl or

NaOH, if necessary. Make up the volume to 1.0 L, sterilize by filtering or autoclaving (20 min, 15 psi, liquid cycle), and store at room temperature.

10. 100× Collagenase–dispase solution: dissolve 100 mg collagenase–dispase powder in 1 mL PBS. Filter-sterilize and store in aliquots at −20 °C.
11. 2× *N*-2-Hydroxyethylpiperazine-*N*′-2-ethanesulfonic acid (HEPES)-buffered saline (HeBS): dissolve 8.2 g NaCl, 5.95 g HEPES, and 105 mg Na_2HPO_4 in 400 mL double-distilled water and adjust pH to exactly 7.05. The exact pH is extremely important for efficient transfection (the optimal pH range is 7.05–7.12). Add water to 500 mL, filter-sterilize, and store in aliquots at −20 °C. Avoid repeated freezing and thawing. Test for transfection efficiency with each new batch (*see* **Note 1**).
12. 2.5 M $CaCl_2$ solution: dissolve 36.74 g $CaCl_2 \cdot 2H_2O$ in 500 mL of double-distilled water. Filter-sterilize and store in aliquots at −20 °C. This solution can be frozen and thawed repeatedly.
13. Antibiotics for selection:

    100× G418 solution: prepare 50 mg/mL active ingredient of G418 in double-distilled water, filter-sterilize, and store at 4 °C protected from light.

    Blasticidin S solution conc. 10 mg/mL.
14. Polybrene solution: dissolve 8 mg polybrene in 1 mL sterile PBS (1,000× stock solution).
15. Crystal violet solution: dissolve 5 mg crystal violet in 125 mL methanol and 375 mL double-distilled water.
16. Tissue culture media (Dulbecco's modified Eagle's medium [DMEM], trypsin–ethylene diamine tetraacetic acid solution).
17. BIO-AMF-2 medium (Biological Industries, Kibbutz Beit Haemek, Israel).
18. For efficient propagation of primary human amniotic fluid cells supplement Ham's F10 medium with 10 % FCS, 1 % Penicillin/Streptomycin, and 1 % Ultroser™ G (Pall GmbH; cat.no. 15950-017).

3 Methods

3.1 Preparation of BRK Cells

1. Anesthetize 5–7-day-old neonatal rats, normally a litter of 6–12 pups, by exposure to isoflurane. Alternatively, kill the animals by cervical dislocation (*see* **Note 2**). Place the pups on a layer of sterile paper towels and remove a section of the skin on the dorsal side (*see* Fig. 1). Open the abdominal cavity by an incision starting from the iliosacral region up to the sternum.

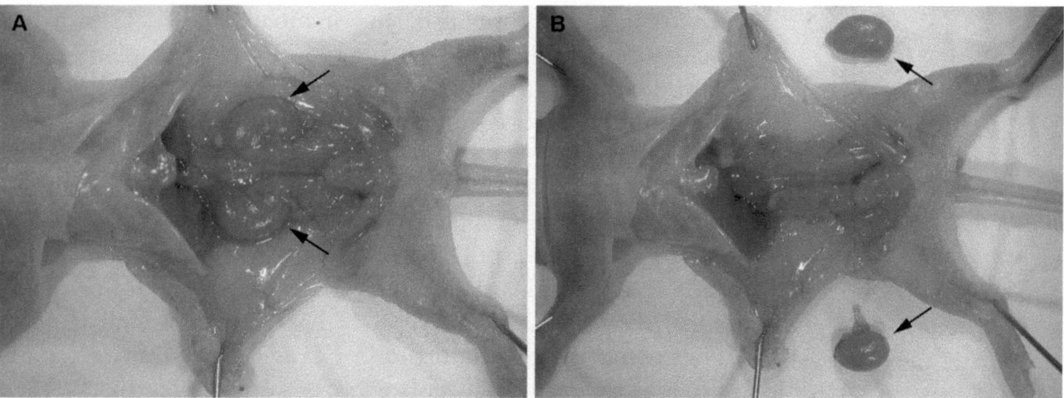

Fig. 1 (**a**) Illustration of the rat abdominal cavity after complete resection of the gastrointestinal tract and the liver. Both kidneys (*arrows*) are clearly exposed and can be removed with sterile forceps (**b**)

Flap skin and adhering abdominal wall. Completely remove the gastrointestinal tract and liver. Remove both kidneys with sterile forceps. Transfer kidneys to a sterile Petri dish or 50-mL tube containing PBS.

2. Wash the kidneys with PBS and transfer to a fresh Petri dish containing a small volume of PBS.
3. Remove capsules and attached blood vessels with pointed forceps (*see* **Note 3**).
4. Transfer kidneys to a new Petri dish containing a small volume of PBS and mince into small pieces (approx 0.1–0.5 mm in diameter) using a sterile scalpel and/or scissors.
5. Transfer minced kidney pieces (~10 kidneys/25 mL PBS) to a 50-mL sterile conical tube. Add collagenase–dispase solution to a final concentration of 1 mg/mL.
6. Vortex tissue suspension and incubate for approx 1 h at 37 °C. Vortex approximately every 15 min for approx 30 s.
7. Wait for 5 min to allow the clumps to settle, and transfer the supernatant to a new 50-mL sterile centrifuge tube kept on ice.
8. Add new PBS (~25 mL/10 kidneys) containing 1 mg/mL collagenase–dispase to the clumps (in original 50-mL tube), vortex, and incubate for another 1–3 h until the clumps are dispersed (*see* **Note 4**).
9. Pool the supernatant (Subheading 3.1., **step** 7) and second digest and centrifuge for 5 min at 800 × *g* and 4 °C.
10. Remove the supernatant and suspend the cell pellet in a small volume of prewarmed (37 °C) DMEM containing 10 % fetal bovine serum and antibiotics.
11. Disperse the cells with the help of a pipet and dispense the amount of cells corresponding to one-half kidney into one

10-cm tissue culture dish, and incubate at 37 °C in a CO_2 incubator overnight. Approximately 24 h after plating, cells can be either transfected with DNA directly or passaged once at a 1:4 ratio before transfection (*see* **Note 5**).

3.2 Transfection of BRK Cells

To analyze the immortalizing and transforming functions of human Ad *E1* and *E4* oncogenes, plasmids expressing the individual (viral) gene products (e.g., pcDNA3 derivatives encoding E1A-12S, E1A-13S, E1B-55 kDa, E1B19kDa, *E4orf3*, or *E4orf6*) or combinations thereof (e.g., pAd5 *Xho*I-C 15 carrying a viral genomic fragment comprising the entire Ad5 *E1* region) are transfected into BRK cells. The use of plasmids carrying a selectable marker gene (e.g., the *neo* marker, which allows for selection with G418) facilitates clearing of nontransfected cells and abortive foci induced by *E1A*-mediated transient cell proliferation, thus markedly reducing background. However, selection for a marker gene eliminates not only transient but also stably transformed foci derived by "hit-and-run" transformation mechanisms that do not require permanent maintenance of oncogenes [10]. For cooperative transformation with strong transforming genes such as combinations of *E1A* with activated *ras*, the use of a dominant selection procedure is not generally needed. We routinely transfect BRK cells using a modified calcium phosphate protocol (*see* **Note 6**). Make sure to prepare duplicate or triplicate samples.

1. Mix 1–5 μg of plasmid DNA containing the transforming gene(s) with carrier DNA to give a total DNA amount of 20 μg. Add 50 μL of 2.5 M $CaCl_2$ solution and make up to 500 μL with sterile, double-distilled water in a 1.5-mL reaction tube. Place 500 μL of 2× HeBS buffer into a 5-mL (12×75 mm) polystyrene round-bottom snap cap tube.
2. Transfer the DNA–$CaCl_2$ mixture drop by drop to the 500 μL of 2× HeBS (with a 1-mL pipet and a plastic tip) while vortexing HeBS at medium speed.
3. Allow precipitate to sit for 20 min at room temperature.
4. Add the total volume (1 mL) of the precipitate to each 10-cm dish of cells containing 10 mL growth medium and mix gently.
5. Incubate at 37 °C in a CO_2 incubator for at least 5 h or overnight.
6. Gently shake the dish to dislodge the residual precipitate and remove the medium. Wash once with 10 mL of PBS, feed with 10 mL of fresh medium, and incubate at 37 °C.
7. Optional: add G418 (500 μg/mL) within the next few days.
8. Change medium every fourth day. In assays involving stronger transforming genes, such as the activated *ras* oncogene, the

transfected cells can be stained 10–14 day after transfection. For weaker cooperating oncogenes, such as Ad E1B, cells can be stained 15–20 day after transfection.

3.3 Isolation and Cultivation of Human Amniotic Fluid Cells (AFC)

Amniotic fluid is a rich source for a mixture of different human cell types. Out of this unspecific cell pool it is easy to isolate plastic adherent cells, which can be further cultivated and propagated. Here we show an easy method to isolate and cultivate AFCs for the study of viral oncogenic transformation.

1. 10–15 mL freshly isolated amniotic fluid was centrifuged for 5 min at $800 \times g$ and RT.
2. Discard the supernatant and resuspend the pellet in 2–3 mL BIO-AMF-2 medium, and plate the whole volume into one well of a 6-well tissue culture plate.
3. 7 days post plating aspirate the medium and feed the plastic adherent cells with 2–3 mL of fresh BIO-AMF-2 medium.
4. If cells are 70–80 % confluent, trypsinize them and passage them to a 10-cm dish containing 8–10 mL Ham's F10 medium supplemented with 10 % FCS, 1 % Penicillin/Streptomycin and 1 % Ultroser™ G.
5. Passage and cultivate cells until they have grown to an appropriate number for your experiments.

3.4 Transduction of AF Cells

To study the growth stimulating and transforming functions of human Ad *E1* and *E4* oncogenes, lentiviral vector constructs expressing the individual (viral) gene products (e.g., LeGO vector derivatives encoding E1A-12/13S, E1B-55 kDa, E1B19kDa, E4orf3, or E4orf6) or combinations thereof are transferred to primary human AF cells (*see* **Note 7**). We routinely transduce AF cells using a modified protocol of Weber et al., 2008 [12]. Make sure to prepare duplicate or triplicate samples.

1. Seed 3×10^4 AF cells per well in 500 μL medium in a 24-well plate.
2. Wait 2–5 h for the cells to attach.
3. Replace medium with 500 μL of medium containing 8 μg/mL Polybrene.
4. Add viral particle containing supernatant to the cells according to an MOI of 0.3 particles per cell.
5. Centrifuge the plate for 1 h, $1{,}000 \times g$ at RT.
6. Change medium the next day. Use 1 mL per well (without Polybrene).
7. Cultivate cells for another 3 days.
8. Transfer cells to 10-cm dishes.

9. For further cultivation use medium supplemented with selection antibiotics.
10. 2–3 weeks post transduction the cells can be stained with crystal violet.

3.5 Staining of Immortalized Colonies and Transformed Foci

The immortalized/transformed colonies become visible approx 1 week after transfection. After periodical visual examination of the colonies, if the sizes of the colonies appear satisfactory, stain with crystal violet (or Giemsa) for quantitation and photography.

1. 1 day prior to staining, feed cells with fresh medium.
2. Remove medium and wash once with 5 mL PBS.
3. Fix and stain the colonies with crystal violet solution for 5 min, and wash two times with tap water.
4. Dry the dishes, count the colonies, and photograph if necessary.
5. The immortalized and transformed colonies obtained with *E1A/E1B* and *E1A/E1B* plus *E4orf6* are shown in Figs. 2 and 3.

3.6 Characterization of Transformed Cells

To determine the extent of transformation, the transformed cells are generally assayed for their ability to form anchorage-independent colonies on semi-solid (soft-agar) medium in vitro and/or examined for their ability to form tumors in immunodeficient mice or immunocompetent syngeneic hosts. Here we describe simple tumorigenicity and malignancy assays in athymic nude mice.

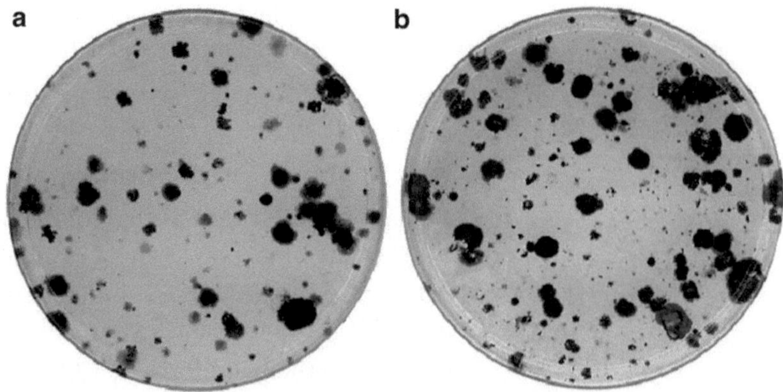

Fig. 2 Focus formation by (**a**) Ad5 *E1A* and *E1B*, or (**b**) *E1A*, *E1B*, and *E4orf6*. Baby rat kidney cells were transfected with 5 μg each of plasmids expressing *E1A/E1B* and *E4orf6* as indicated using the calcium phosphate precipitation technique. Transfected cells were stained with crystal violet 27 days after transfection and scanned

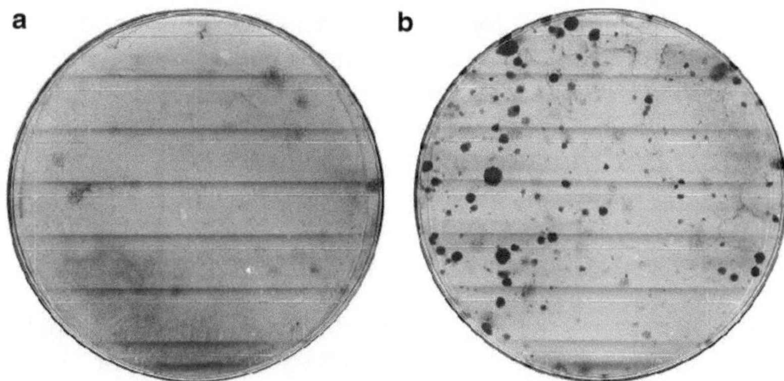

Fig. 3 Focus formation by Ad5 *E1A* and *E1B*. Primary human amniotic fluid cells were transduced with (**a**) empty LeGO vector particles, or (**b**) co-transduced with Ad5 *E1A* and *E1B* harboring LeGO vector particles with an MOI of 0.3 particles per cell. Transduced cells were stained with crystal violet 27 days after transfection and scanned

3.6.1 Tumor Incidence and Tumor Growth in Athymic Mice

1. Grow cells on 10- or 15-cm tissue culture dishes. At the day of injection, dishes should contain 50–80 % confluent monolayers. Each transformed cell line should be tested in at least five mice, and for each animal 1×10^6 cells are needed.
2. Wash monolayers two times with serum-free DMEM and scrape cells into serum-free medium using a sterile plastic cell lifter. Carefully transfer cell suspension to a conical tube and centrifuge for 5 min at $800 \times g$.
3. Remove supernatant and add serum-free medium to make up a suspension that contains 10^7 cells/mL (1×10^6 cells per injection).
4. Inject 100 μL of cell suspension into nude mice subcutaneously using a syringe equipped with a 26-G needle.
5. Examine the site of injection weekly for development of tumor nodules to determine the latency period. After development of visible (palpable) tumors, measure the length and width of the tumors and calculate the approximate tumor area. Results are displayed as tumor incidence (no. of mice with tumors/total no. of animals) and tumor size (mean tumor area ± standard errors [mm^2]).

3.6.2 Tumor Histology and Metastasis Assay

For histological analysis, tumors as well as surrounding connective tissue and adhering skin regions are fixed in Bouin's solution and prepared for routine paraffin histology. Sections (5 μm) are stained with hematoxylin–eosin according to the Masson and Golder method modified by Jerusalem. Athymic mice injected with transformed cells that induce tumors can also be examined for formation of metastatic tumor nodules in the internal organs such as lungs or lymph nodes. If the burden of primary tumors is heavy for the animals (tumor area ≥100 mm^2), the primary tumors must be surgically removed, after which the animals are maintained for an additional 2–4 weeks and autopsied.

4 Notes

1. The 2× HeBS solution can be rapidly tested by mixing 0.5 mL of 2× HeBS with 0.5 mL of 250 mM $CaCl_2$ and vortexing. A fine precipitate should develop that is readily visible in the microscope.
2. Neonatal Fisher rats are equally suitable. Kidneys may also be obtained from animals between 2 and 4 days after birth, which, however, will result in reduced tissue yields.
3. During preparation of BRK cells, as much of the capsule present in the kidney should be removed as possible. If it is not fully removed, the preparation will contain fibroblasts. Fibroblasts will grow faster than epithelial cells, causing increased background.
4. Try to minimize incubation periods because prolonged digestion of minced kidneys with collagenase–dispase will result in reduced viability and reduced plating efficiency of epithelial cells.
5. Continued passaging before transfection will result in decreased transformation efficiencies.
6. Other transfection methods including liposome-mediated techniques may also be employed. BRK cells are also highly suited for gene transduction with adequate retroviral vectors.
7. The use of lentiviral vectors with a selectable marker gene (e.g., the neo and bsr marker, which allows for selection with G418 and blasticidin S) facilitates clearing of non-transduced cells and abortive foci induced by E1A-mediated transient cell proliferation, thus markedly reducing background.

References

1. Endter C, Dobner T (2004) Cell transformation by human adenoviruses. Curr Top Microbiol Immunol 273:163–214
2. Williams JF, Zhang Y, Williams MA, Hou S, Kushner D, Ricciardi RP (2004) E1A-based determinants of oncogenicity in human adenovirus groups A and C. Curr Top Microbiol Immunol 273:245–288
3. Chinnadurai G (2004) Modulation of oncogenic transformation by the human adenovirus E1A C-terminal region. Curr Top Microbiol Immunol 273:139–161
4. White E (2001) Regulation of the cell cycle and apoptosis by the oncogenes of adenovirus. Oncogene 20:7836–7846
5. Täuber B, Dobner T (2001) Adenovirus early E4 genes in viral oncogenesis. Oncogene 20:7847–7854
6. Glaunsinger BA, Weiss RS, Lee SS, Javier RT (2001) Link of the unique oncogenic properties

of adenovirus type 9 E4-ORF1 to a select interaction with the candidate tumor suppressor protein ZO-2. EMBO J 20:5578–5586
7. Thomas DL, Schaack J, Vogel H, Javier RT (2001) Several E4 region functions influence mammary tumorigenesis by human adenovirus type 9. J Virol 75:557–568
8. Nevels M, Rubenwolf S, Spruss T, Wolf H, Dobner T (2000) Two distinct activities contribute to the oncogenic potential of the adenovirus type 5 E4orf6 protein. J Virol 74:5168–5181
9. Nevels M, Täuber B, Kremmer E, Spruss T, Wolf H, Dobner T (1999) Transforming potential of the adenovirus type 5 E4orf3 protein. J Virol 73:1591–1600
10. Nevels M, Täuber B, Spruss T, Wolf H, Dobner T (2001) "Hit-and-run" transformation by adenovirus oncogenes. J Virol 75:3089–3094
11. Schiedner G, Hertel S, Kochanek S (2000) Efficient transformation of primary human amniocytes by E1 functions of Ad5: generation of new cell lines for adenoviral vector production. Hum Gene Ther 11:2105–2116
12. Weber K, Bartsch U, Stocking C, Fehse B (2008) A multicolor panel of novel lentiviral "gene ontology" (LeGO) vectors for functional gene analysis. Mol Ther 16:698–706

Chapter 9

Oncolytic Adenovirus Characterization: Activity and Immune Responses

Raul Gil-Hoyos, Juan Miguel-Camacho, and Ramon Alemany

Abstract

Virotherapy in one of the main current applications of recombinant adenoviruses. Oncolytic adenovirus are designed to target tumors, replicate selectively in tumor cells, and elicit immune responses against tumor antigens. Transgene expression in replication-competent oncolytic vectors allows to explore multiple strategies to enhance the potential of virotherapy. In this chapter we describe common in vivo and in vitro techniques used to evaluate the potency and biodistribution of oncolytic viruses. Monitoring immune responses against viral and tumor antigens is crucial as the immune system determines the outcome of virotherapy.

Key words Adenovirus, Oncolytic, Virotherapy, Cancer, Immune responses

1 Introduction

During recent years, many strategies have been used to increase the potency of oncolytic adenoviruses. Arming adenoviruses with exogenous genes encoding suicide or prodrug-activating proteins, cytotoxic proteins, proteases, or immunomudulatory cytokines has improved the prospects of oncolytic adenoviruses [1, 2]. There are different strategies for the expression of the exogenous genes from the viral genome. The transgene can be expressed using the endogenous E1a, E3, or major late promoters, linked to these expression units with internal ribosome entry sites (IRES) or splicing acceptors. This strategy is advantageous compared to the use of exogenous promoters (CMV and other promoters) because adenovirus (commonly used serotype 5) can package up to 38 kb of DNA, which is only 2 kb over the genome size. This packaging limitation is important if we consider that most oncolytic adenoviruses also contain tumor-selective promoters that regulate viral genes, commonly the E1a transcription unit. The deletion of some E3 genes (those associated to immune-inhibitory functions but not the

Miguel Chillón and Assumpció Bosch (eds.), *Adenovirus: Methods and Protocols*, Methods in Molecular Biology, vol. 1089, DOI 10.1007/978-1-62703-679-5_9, © Springer Science+Business Media, LLC 2014

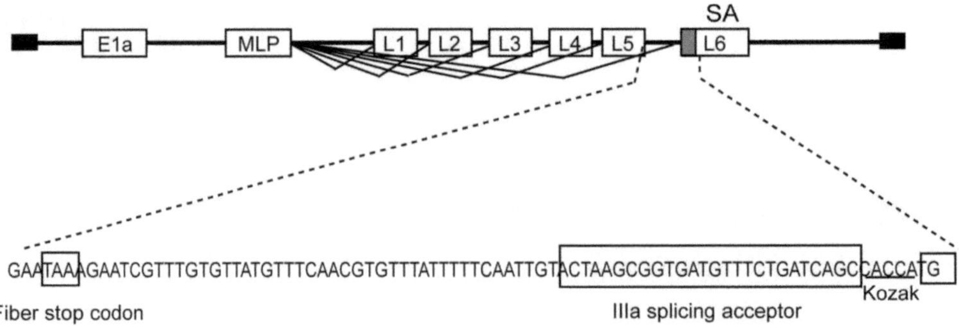

Fig. 1 IIIa splicing acceptor gene expression system. A transgene is expressed as a new late transcriptional unit regulated by the MLP. The splicing acceptor from the adenovirus gene IIIa is located right after the fiber polyA

adenovirus death protein) is an option to create space for exogenous transgenes. However, this E3 deletion may accelerate virus clearance in an immunocompetent host [3]. As splicing acceptors are much shorter (26bp) compared to IRES (500 bp), they represent our preferred option to arm adenoviruses with transgenes. Another option that also saves cloning space for transgenes is to link the transgene to a virus protein using the 2A sequence that during translation precludes the formation of the peptide link [4] and produces two separate proteins from the same mRNA.

We have used the adenoviral splicing acceptor IIIa after the fiber open reading frame to express transgenes. This acceptor creates a new splicing unit in the long major late transcript (which would correspond to L6), without affecting the virus cycle [5, 6]. We also usually insert the Kozak sequence (CCACC) before the ATG of the start codon of the gene of interest to improve translation initiation. Finally, we use the TAA stop codon to terminate the transgene and add TAAA after it to form a polyA signal (*see* Fig. 1). Late gene expression is an advantage when the proteins encoded by the gene of interest can interfere with the viral cycle, such as suicide, proapoptotic, or cytotoxic proteins.

In this chapter we describe basic protocols for the characterization of oncolytic adenoviruses. We have not included virus construction and purification protocols as those are described in other chapters of this book. The infectious unit titer and the IC50 assays are used to measure the concentration of active virus in a sample. A differential infectious unit titer in two cell lines can be used to assess the relative permissiveness of a given virus in these two cell types. If a given cell line is infected with several viruses at the same multiplicity of infection (MOI or infectious units/cell) or with the same physical particle titer (vp/cell), the IC50 reflects the propagation efficacy (in vitro oncolytic potency). The size of a plaque may also indicates this propagation efficiency, but it is less quantitative and more dependent on the cell line used, because cells that do not detach easily do not form clear plaques. In vivo, the follow up of tumor size after intravenous or intratumoral administration of the oncolytic adenovirus is commonly used to define potency. Histology combined with quantitative information of virus DNA content and distribution of the virus in the tumor (protein staining) are used to correlate antitumor activity with biological parameters such as the amount of virus that reaches the tumor (targeting), the amount of virus replication and spread, and the blocking effect of stroma barriers and immune responses. Paraffin embedding preserves better the morphology of tissues but if antibody fails to detect the antigen, staining on frozen sections is required. Therefore, it is recommended to keep tissues for both protocols.

Adenovirus is a very strong inducer of antibody and cellular immune responses. We describe a quick method to measure neutralizing antibodies against adenovirus. Specific anti-adenovirus CTL responses have been detected in C57BL/6 mice (H2Db-restricted) using peptides from E1a (SGPSNTPPEI) [7], E1b (VNIRNCCYI) [8], and hexon (GGVINTETL and FAIKNLLLL) (Gil-Hoyos unpublished data). ELISpot and intracellular cytokine staining (ICS) are good techniques to monitor cellular immune responses suing splenocytes isolated from treated mice. In vivo killing assays are appropriate to test these cellular immune responses in vivo. For this assay splenocytes are incubated with the antigen or epitope peptide and labeled with a high concentration of the fluorescent marker CFSE. These high-CFSE splenocytes and an equivalent amount of splenocytes control (without antigen or non-pulsed) labeled with low-CFSE are injected in the animals where specific CTL immunity wants to be determined. The lysis of these target cells relative to the lysis of nontarget cells indicates the CTL activity against the antigen or the peptide.

2 Materials

2.1 Titration

1. Dulbecco's Modified Eagle's Medium (DMEM) supplemented with 10 % fetal bovine serum.
2. PBS-Ca/Mg: 10 g/L $MgCl_2$ Solution (2.5 g $MgCl_2$ hexahydrate in 250 mL H_2O, filter sterilize). 10 g/L $CaCl_2$ Solution (2.5 g $CaCl_2$ dihydrate in 250 mL H_2O, filter sterilize). PBS 10× (sterile). In a 1 L beaker, add 100 mL PBS 10×, 10 mL MgCl2 solution, 10 mL $CaCl_2$ solution, and cold autoclaved H_2O up to 1 L.
3. BCA protein stain: mix one part of reactive A in 50 parts of reactive B and vortex the solution. This solution can be stored for 24 h if required. For 96 wells: 96 wells × 200 μL/well = 19.2 mL reactive A + B => 20 mL reactive A + 400 μL reactive B.

2.2 Plaque Assay

1. Neutral Red staining (classic method): 0.03 % Neutral red solution in PBS. Vortex vigorously to dissolve well as neutral red forms crystals if not dissolved properly.
2. Thiazolyl Blue Tetrazolium Bromid (MTT) staining: 5 mg/mL stock solution of MTT in PBS or DMEM 5 % FBS.

2.3 qPCR

1. Takara PreMix 2× (cat #: RR039A).
2. Hexonprimersandprobe: Ad18852 (sequence 5′-CTTCGATGA TGCCGCAGT G-3′) stock 10 μM; Ad18918R (sequence 5′-GGGCTCAGGTAC TCCGAGG-3′) stock 10 μM, probe 5′-FAM-TTACATGCACATCTCGGGCCAGGAC-TAMRA-3′. stock 10 μM.

2.4 In Vivo Studies

1. Tissue adhesive: Vetbond 3 M, 3 M Animal Care Products, St. Paul. MN.
2. Primary anti-adenovirus antibody: Rabbit polyclonal Abcam ref. Ab6982.
3. Secondary antibody: Envision System. Labeled Polymer-HRP anti-rabbit, DAKO Cytomation.
4. 10× citrate sodium solution: For 500 mL mix 1.9 g citric acid in 90 mL sterile water and 12.05 g sodium citrate in 410 mL sterile water.
5. DAB (3,3′ Diaminobenzidine) buffer: 1,000 μL PBS + 4 μL hydrogen peroxide.
6. DAB staining mix: 20 μL of DAB chromogen (DAKO) solution mixed with 1 mL of DAB buffer.
7. ACK lysis buffer: 150 mM NH_4Cl, 10 mM $KHCO_3$, 0.1 mM Na_2EDTA. Adjust to pH 7.2–7.4 with 1 N HCl. Filter sterilize and store at 4 °C.
8. FACS Buffer: PBS 5 % FBS, 0.5 % BSA, 0.07 % NaN_3 (Sodium azide). Filtered with 0.22 μm filter.
9. Perm Wash (permebilization): PBS with 1 % FCS, 0.1 % NaN_3, 0.1 % Saponine. Filtered with 0.22 μm filter.
10. FACS Fixative: 0.89 % NaCl, 1 % PFA. Adjust to pH 7.4 with NaOH.
11. Intracellular staining first antibody mix:

    CD8 PECy5: 1.5 μL of 1:10 dilution + 48.5 μL of FACS buffer.

    CD4 PerCPCy5.5: 1.5 μL of 1:10 dilution + 48.5 μL of FACS buffer.
12. Intracellular staining second antibody mix:

    IFNg APC: 1.5 μL of 1:10 dilution + 48.5 μL of Perm Wash.

13. ELISpot coating antibody solution: 4 μg/mL Anti-IFNg AN18 diluted in PBS.
14. Detection antibody solution: 1 μg/mL Anti-IFNg R4-6A2-biotin diluted in PBS 0.5 % FBS.
15. Streptavidin-Alkaline Phosphatase (ALP) solution (Sigma): dilute 1:1,000 in PBS 0.5 % FBS.
16. BCIP/NBT solution (Sigma).
17. CFSE stock solution: Dissolve 5-(6-) carboxyfluorescein diacetate succinimidyl ester (CFSE; Sigma ref. 21888) in 5 mM dimethylsulfoxide. Store aliquots at −20 °C. The stock solution can be refrozen quickly after thawing. Prepare the diluted CFSE just before use and do not freeze it.

3 Methods

3.1 Adenovirus Titration by Anti-Ad Staining (Infectious Unit Titer)

1. In a 96-well dish prepare 1/10 serial dilutions of the virus stock or sample using DMEM/5 % FBS in a final volume of 100 μL. Add 100 μL of medium to all wells needed and then add 10 μL of concentrated virus to the first column. Use three lanes per sample (triplicate). Change tip every time after the virus is taken to the next column and mixed. Prepare at least ten serial dilutions, although this range can be adjusted considering an approximate titer of 10^8 infectious units (iu)/mL in a cell extract and 10^{10} iu/mL in a purified virus. Do not add virus to the last column (negative control) (*see* **Note 1**).
2. Prepare a cell suspension of 10^6 cells/mL in DMEM + 5 % FBS and add 50 μL to each well (50,000 cells/well). Select a proper cell line according to the characteristics of the virus to analyze (typically HEK293 cells).
3. Incubate the infected cells at 37 °C for 24–36 h (for HEK293 cells) (*see* **Note 2**).
4. Carefully remove the medium with a tip connected to vacuum pump or to a pipette. Avoid cells loss. Let the cells air-dry for 10 min.
5. Fix cells with 100 μL/well of 100 % methanol for 10 min at −20 °C (this step can be extended for the desired time).
6. Prepare the primary antibody (anti-hexon, anti-fiber, or anti-adenovirus polyclonal) in PBS-Ca/Mg, 1 % BSA. For a hybridoma supernatant (we use 2H×-2 hybridoma) prepare a 1/5 solution. For a purified anti-adenovirus antibody prepare a 1/500 dilutions.
7. Remove the methanol from the titrating plate without contacting the monolayer.

8. Wash each well twice with 100 μL PBS-Ca/Mg, 1 % BSA (ions prevent cells detachment but PBS without ions can be also used). During washing steps it is very important avoid air-dry the cells.
9. Remove washing solution and add 50 μL of the diluted Ab to each well.
10. Incubate for 1–2 h at 37 °C.
11. Wash three times each well with 100 μL of PBS-Ca/Mg 1 % BSA.
12. Dilute the secondary antibody (FITC- or Alexa488-conjugated) in PBS-Ca/Mg, 1 % BSA. For most commercial antibodies dilution 1/300 is recommended. Remove washing solution and add 50 μL of the dilution to each well and incubate 1–2 h at 37 °C in dark.
13. Wash three times each well with 100 μL of PBS-Ca/Mg, 1 % BSA. Try to avoid long exposition to light.
14. In an inverted fluorescence microscope quantify the number of green cells in each well (*see* **Note 3**). Concentration will the mean of the triplicate multiplied by the dilution factor. Note that 100 μL of each dilution were used so a tenfold factor must be applied to obtain the iu/mL titer.

3.2 Ad Bioactivity Titration Assay (IC_{50})

1. Tripsinize and prepare a 3×10^5 cells/mL suspension in DMEM 5 % FBS of HEK293 or the cell line tested for permissiveness.
2. Prepare serial dilutions of the virus. For cells that are easily infected use 1/5 dilutions. For cells that need high MOI to be infected use 1/2 or 1/3 serial dilutions. Prepare 11 serial dilutions with DMEM/5 % FBS of each virus in a 96-well plate with a final volume of 50 μL/well. Change pipette tips every time after virus is taken and mixed in the next well.
3. In the last column (column 12) add a volume of 50 μL/well of DMEM/5 % FBS but no virus. This will be the negative control.
4. Decide the highest vp/cell or iu/cell (MOI) desired to infect the cells in first column in order to obtain complete cytopathic effect (CPE) in 6–10 days. For example 5333.3 vp/cell or 100–300 iu/cell. The vp/mL needed to infect the first column is

   5333.3 vp/cell × 30,000 cell/well = 1.5×10^8 vp/50 μL = 3×10^6 vp/μL.
5. Dilute the viruses in order to obtain 3×10^6 vp/μL. Prepare a serial dilution by transferring either 12.5 μL (1/5 dilution) or 25 μL (1/3 dilution) to 50 μL of DMEM 5 % FBS in the next column of wells. Leave the last column without virus (negative control).
6. Add 100 μL of cell suspension to each well, from the negative control column to the most concentrated.

7. Incubate at 37 °C 5 % CO_2 in incubator and observe daily under the microscope the appearance of CPE. Leave until partial CPE can is seen in the sixth or seventh column (5–8 days post-infection). First column should show complete CPE.
8. Prepare BCA protein stain.
9. Shake softly the 96-well plate in order to resuspend detached cells and aspirate the medium from every well with care. Wash gently cells with 200 μL of PBS to eliminate the remaining FBS that would mask the differences in protein content.
10. Remove the PBS and quickly add 200 μL/well of BCA protein stain with a multichannel pipette.
11. Leave 30 min at 37 °C and read absorbance in spectrophotometer at 540 nm. Use as blank any well from first column.
12. Record results in an excel file and plot the amount of protein as a percentage of the noninfected control against vp/mL or iu/mL.
13. Calculate the IC50 for each virus by nonlinear regression using an adapted Hill equation (e.g., GraFit software, Erithacus, Horley, UK).

3.3 Plaque Size Assay

1. Seed HEK293, A549, or the desired cell line in 6-well plates in order to have 90 % confluence in the time of infection. Before the infection, remove the medium and add 0.9 mL of DMEM 5 % FBS to each well.
2. Prepare 1/10 dilutions of the virus stock or cell extract (CE) and add 100 μL of each dilution per well. Proceed from the most diluted to the most concentrated in order to use the same pipette tip.
3. Incubate at 37 °C/5 % CO_2 for 4 h.
4. Prepare autoclaved 1 % agarose in H_2O and temperate to 50 °C (previously prepared agarose can be melted in a microwave).
5. 4 h after infection, remove infection medium and wash cells with PBS (*see* **Note 4**). Remove medium and PBS starting from the diluted wells to prevent contamination. Handle one plate at a time.
6. Mix 7.5 mL of 1 % agarose (at 50 °C) and 7.5 mL of DMEM 5 % FBS (at 37 °C) in a falcon tube. Remove the PBS and add 2 mL of the agarose-medium mix in each well. Leave 10 min on the hood without lid. Repeat this overlay for each plate.
7. Add 2 mL of DMEM 5 % FBS to each well and transfer to the cell incubator.
8. Incubate at 37 °C and 5 % CO_2 until plaques appear. If the medium acidifies (turns yellow) replace it carefully.
9. Stain plaques with neutral red adding 1 mL of 0.03 % neutral red into each well and incubating O/N at 37 °C and 5 % CO_2.

Remove neutral red and place on a white transilluminator to capture images (e.g., GelDoc imaging system) (*see* **Note 5**).

3.4 Antitumor Efficacy In Vivo

1. To implant tumors prepare a suspension of exponentially growing cells (subconfluent plates). The cells implanted to generate tumors vary from 10^6 to 10^7 according to every cell line.
2. Inject 0.15 mL of cell suspension in each flank of the animal (mouse or hamster) with a 29 G needle.
3. Follow up animals every 3 days and measure length (A) and width (B) of each tumor with a caliper. Volume $(\text{mm}^3) = A \times B \times (\pi/6)$. Once the tumors reach 100–150 mm^3, animals must be distributed randomly in groups.
4. Dilute purified adenovirus in PBS for injection. For intravenous or intraperitoneal administration use a volume of 200 μL (typically we use 5×10^{10} or 10^{11} vp in a 25 g mouse and 4×10^{11} in a 100 g hamster). Inject in the lateral tail veins rubbing them with alcohol for dilation, using a 1 mL syringe with a 30 G needle. For intratumoral injection used anesthesia, inject a maximum of 20 μL/tumor using small syringes (0.3 mL) and 30 G needles, and distribute this volume is multiple (2–4) injections. Wait for 1 min before removing the needle and between injections. To avoid the leakage of the injected virus upon sequential intratumoral injections it is recommended to seal the injection puncture with tissue adhesive.
5. Follow tumor volume with a caliper every 2 or 3 days. Animals must be handled according animal use and care committee recommendations, usually avoiding ulceration and size over 1 mL.

3.5 Adenovirus Genome Biodistribution by qPCR

1. Sacrifice animals (CO_2 chamber) and resect organ (tumor or others) into eppendorf tubes to snap freeze in liquid nitrogen or in dry ice/ethanol. Homogenize the tissue in liquid nitrogen with a pestle and mortar on dry ice until a powder is obtained. Extract total DNA using commercial kits (e.g., NucleoSpin DNA Blood, Macherey-Nagel; QiAamp DNA Mini, Qiagen).
2. Aliquots of resuspended homogenate before lysis from a negative tumor or tissue (vehicle control animals) can be spiked with nine serial dilutions of purified virus of known concentration (from 10 to 10^9 vp/μL) to generate a standard curve.
3. Dilute DNAs to a final concentration of 25 ng/mL.
4. Mix 4 μL of each DNA sample (100 ng), with 0.3 μL of 10 μM hexon primers, 0.1 μL probe, 5 μL of 2× Premier Extaq, and 0.3 μL of H_2O.
5. Amplify in LightCycler (Roche) heating at 10 min a 95 °C followed by 40 cycles of 95 °C—15 s/60 °C—1 min, to obtain the amplification curves of the standard curve and samples.

6. Analyze the amplification curves (threshold cycle to obtain positive signal or crossing point) to obtain de virus DNA concentration of the sample.

3.6 Immunohistochemistry with Frozen Tissue Sections

1. Embed fresh tissues carefully in OCT in plastic cryomold, taking care not to trap air bubbles surrounding the tissue. Freeze tissue by setting the mold on dry ice. The frozen blocks may be stored at −80 °C for a long term.
2. Before cutting sections, allow frozen blocks to equilibrate at −25 °C in the cryostat chamber for about 5 min.
3. Cut 5 μm sections on slides, placing two or three separated sections per slide. For better adhesion use poly-l-lysine slides. Slides can be stored at 4 °C for a few days or at −80 °C long term.
4. To stain, let the sections dry for at least 30 min at RT and fix them in 4 % formaldehyde for 10 min at RT.
5. Rinse sections three times for 2 min in deionized water and once for 10 min with PBS + 0,05 % Tween 20.
6. Carefully dry slides with paper and encircle sections using a water repellent marker (PAP PEN).
7. Block sections with 20 % goat or horse serum in PBS for 30 min at RT in a sealed humidity chamber.
8. Remove the serum and cover sections with primary anti-adenovirus antibody (e.g., rabbit polyclonal Ab6982) diluted 1/100 in PBS + 0.05 % Tween 20. Incubate 1 h at RT in a humidity chamber. For a negative control incubate one section of each slide with PBS or with a unspecific rabbit antibody diluted 1/100 in PBS + 0.05 % Tween 20.
9. Wash the sections three times for 5 min in PBS + 0.05 % Tween 20.
10. Without letting sections dry out, add the secondary antibody (e.g., anti-rabbit labeled with fluorescence) diluted according the provider in PBS + 0.05 % Tween 20. Incubate as before for at least 1 h.
11. Wash the sections three times for 5 min in PBS + 0.05 % Tween 20.
12. Mount coverslips with mounting media, DAPI, and antifading agents, and observed the stained sections in a fluorescence microscope.

3.7 Immunohistochemistry on Paraffin-Embedded Tissue Sections

1. Slice (3 mm width) tumor and organs and place them inside tissue cassettes. Fix tissue in buffered formalin or fresh 4 % paraformaldehyde during 48 h at RT. Dehydrate and embed in paraffin (70 % ethanol twice 1 h; 80 % ethanol twice 1 h; 95 % ethanol twice 1 h, 100 % ethanol, three incubations of 1 h; xylene three incubations of 1 h; and paraffin wax at 58 °C two changes of 1 h.).

2. Cool down paraffin blocks at −20 °C and cut sections of 4 μm of thickness. Once cut, sections are extended in distilled water at 42 °C. For better adhesion use poly-l-lysine slides and deionized water in the water. Fish the slices on the slides and dry at 37 °C for 30 min. Put two paraffin sections per slide in order to have a control without the specific antibody.
3. Deparaffinize and hydrate tissue sections as follows: heat slides for 2 h at 65 °C or overnight at 37 °C. Wash four times for 5 min in xylene; three times for 5 min in 100 % ethanol; three times for 5 min in 96 % ethanol; and once for 5 min in 70 % ethanol. Rinse sections once for 5 min in deionized water.
4. Block endogenous peroxidase incubating sections for 10 min in 0.33 % hydrogen peroxide. Rinse once for 5 min in deionized water and three times for 5 min in PBS pH 7.4.
5. Antigen retrieval: in a pressure cooker submerge slides in 1× citrate sodium solution prepared in PBS and adjusted to pH = 6. Heat to reach maximum pressure, wait 2 min, and stop heating. Wait about 20 min until pressure cooker cools down. Rinse slides once for 5 min in deionized water and three times for 5 min in PBS pH 7.4.
6. Block sections with 20 % goat or horse serum in PBS for 30 min at RT in a sealed humidity chamber.
7. Remove the serum and cover sections with primary anti-adenovirus antibody (e.g., rabbit polyclonal) diluted 1/100 in PBS + 0.05% Tween 20. Incubate 1 h at RT in a humidity chamber. For a negative control incubate one section of each slide with PBS or with a unspecific rabbit antibody diluted 1/100 in PBS + 0.05 % Tween 20.
8. Wash the sections three times for 5 min in PBS + 0.05 % Tween 20.
9. Without letting sections dry out, add the secondary antibody (HRP-conjugated anti-mouse or anti-rabbit) and incubate for 30 min at RT in a humidity chamber.
10. Rinse sections three times for 5 min in PBS + 0.2 % Triton X-100.
11. Cover sections with DAB staining mix and incubate until a brown precipitate appears (2–5 min). Stop reaction by rinsing sections with tap water.
12. Counterstain with hematoxylin for 1 or 2 min. Rinse slides about 10 min with tap water.
13. Dehydrate once for 2 min in 70 % ethanol, three times for 2 min in 96 % ethanol, three times for 2 min in 100 % ethanol, and four times for 5 min in xylene.
14. Mount slides with DPX.

3.8 Neutralizing Anti-Adenovirus Antibody (Nab) Assay

1. Collect blood from tail vein in capillary tubes without heparin or EDTA. Incubate 5 min at RT and spin at 10,000 × *g*, 4 °C to obtain the supernatant (test serum).
2. Transfer the test serum to an eppendorf and heat-inactivate it at 56 °C for 30 min.
3. Using DMEM/5 % FBS, dilute an adenovirus from a purified stock of known physical (vp) or infectious titer to obtain 10^7 vp/mL or 5×10^5 iu/mL (5×10^5 vp/50 μL or 2.5×10^4 iu/50 μL). Each serum to be tittered (in triplicate) requires 1.8 mL of diluted virus (*see* **Note 6**).
4. In a 96-well plate add 90 μL of diluted virus to the first column of wells and 50 μL to the rest except for column 12 (no virus control). In the first row or column (dilution 1/10) add 10 μL of the test serum to the 90 μL of diluted virus. Then perform a ½ serial dilution transferring 50 μL of this row to the next row and successively until row 10. Discard the final 50 μL taken from row 10. Column 11 is no test serum or virus only control. Add 50 μL of DMEM/5 % FBS in well 12 (no virus control).
5. Incubate the virus and test serum (containing the Nab) for 1 h at 37 °C.
6. Prepare a cell suspension (HEK293 cells) of 10^6 cells/mL in DMEM/5 % FBS. Add 100 μL/well (100,000 cells/well). Incubate at 37 °C for 24 h (*see* **Note 7**).
7. Determine titer by anti-hexon staining as in protocol 3.1 above, from **step 4** on. The neutralizing antibody titer is the inverse of the dilution that reduces in 50 % the titer obtained in well 11 (without Nab) (e.g., if 50 % of titer decrease is found in well 6, Nab titer is 320).

3.9 CTL Responses by Intracellular Cytokine Staining

1. Harvest the spleen of the animal to be tested (treated with adenovirus) and leave it in a 15 mL tube in 5 mL of RPMI medium.
2. Put the spleen in a cell strainer and mechanically disrupt the spleen with the flat portion of a syringe plunger until no fragment remains. Transfer the resuspended cells to a 15 mL tube. Rinse the cell strainer with 5 mL of RPMI and transfer to tube. Repeat this strainer wash to harvest all cells. Discard the strainer.
3. Centrifuge the disrupted cells at 500 × *g* for 5–10 min.
4. Remove the supernatant and resuspend the pellet in 2 mL of ACK lysis buffer to lyse red blood cells. Let the cell suspension at RT for 3 min.
5. Fill the tube with RPMI and centrifuge the cells at 500 × *g* for 5 min.

6. Remove the supernatant and resuspend the cell pellet in PBS. Wash the pellet two more times with PBS, discarding tissue debris.
7. Resuspend the cells in 5 mL of RPMI medium.
8. Refilter the cells through a cell strainer if clumps are present.
9. Count the mononuclear cells by trypan blue exclusion using a hemocytometer.
10. Resuspend cells at 10^7 in RPMI.
11. Seed 100 μL/well of the cell dilution (10^6 cells/well) in a 96-well plate with U-bottom.
12. Add 100 μL of stimulant (final volume 200 μL/well): (a) PMA and Ionomycin as positive control. Final concentration: 30 ng/mL PMA + 500 ng/mL Ionomycin; (b) Test peptides (CTL epitopes) at 2 μM final concentration, and (c) RPMI medium (or irrelevant peptide) as negative control.
13. Incubate cells at 37 °C 5 % CO_2 for 2 h to O/N.
14. Add Brefeldin A in the well to a final concentration of 10 μg/mL. Mix gently with the pipette. Incubate at 37 °C 5 % CO_2 for 4 h.
15. Spin the plate at 400 × *g* 4 °C for 5 min. Check visually if there is pellet.
16. Remove the SN by flicking the plates and then blotting once on clean paper. Check visually if there is pellet.
17. Vortex (check that pellet is well resuspended).
18. Add 50 μL/well of antibody mix. Resuspend cells gently by pipetting.
19. Incubate 30 min on ice, dark (cover the plates with aluminum foil). Alternatively cells can be incubated O/N in the fridge.
20. Add 150 μL/well of FACS Buffer. Spin the plate at 400 × *g* 4 °C for 5 min. Remove the SN. Vortex. Repeat this washing step two additional times.
21. Permeabilize cells adding 100 μL/well of 2 % formaldehyde in cold PBS. Resuspend cells gently by pipetting.
22. Incubate 20 min on ice.
23. Add 100 μL/well of Perm Wash. Spin the plate at 400 × *g* 4 °C for 5 min. Remove the SN. Vortex.
24. Add 200 μL/well of Perm Wash. Spin the plate at 400 × *g* 4 °C for 5 min. Remove the SN. Vortex. Repeat this washing step.
25. Add 50 μL/well of second antibody mix (prepared in Perm Wash). Resuspend cells gently by pipetting.
26. Incubate 30 min on ice, dark.
27. Add 150 μL/well of Perm Wash. Spin the plate at 400 × *g* 4 °C for 5 min. Remove the SN. Vortex.

28. Add 200 μL/well of Perm Wash. Spin the plate at 400 × *g* 4 °C for 5 min. Remove the SN. Vortex. Repeat this wash.
29. Resuspend cells in 100 μL/well of FACS Fixative, transfer samples to FACS tubes, fill up to 300 μL with FACS Fixative.
30. Keep at 4 °C, dark, and measure fluorescence staining by FACS within 48 h.

3.10 CTL Responses by ELISPOT

1. Wash an ELIspot 96-well plate five times with 200 μL/well of sterile water (*see* **Note 8**).
2. Add 100 μL/well of coating antibody solution and incubate overnight at 4 °C.
3. Isolate splenocytes following **steps 1–10** of Subheading 3.9.
4. Remove coating antibody and wash with 200 μL/well of PBS.
5. Block membranes with 200 μL/well of RPMI/10 % FBS. Incubate at least for 30 min at RT.
6. Dilute splenocytes at 2.5×10^6/mL. Add 100 μL of cells/well.
7. Add 100 μL of stimulants (final volume 200 μL/well): (a) PMA + Ionomycin as positive control. Final concentration: 15 ng/mL PMA + 250 ng/mL Ionomycin; (b) test peptides (CTL epitopes) at 2 μM final concentration; and (c) RPMI medium (or irrelevant peptide) as negative control.
8. Incubate at 37 °C 5 % CO_2 at least 16 h. Do not move the plate during this time.
9. Remove the cells and wash five times with 200 μL/well of PBS.
10. Add 100 μL/well of detection antibody solution and incubate 2 h at RT.
11. Wash five times with 200 μL/well of PBS.
12. Add 100 μL/well of Streptavidin-ALP solution and incubate 1 h at RT.
13. Wash five times with 200 μL/well of PBS.
14. Add 50 μL/well of BCIP/NBT solution and develop until distinct spots emerge (10–30 min).
15. Stop color development by washing with tap water.
16. Leave the plate to dry. Read spots with an ELIspot plate reader.

3.11 In Vivo Cytotoxicity Assay

1. Isolate splenocytes following **steps 1–10** of Subheading 3.9.
2. Divide cells into two tubes: "peptide-pulsed" and "non-pulsed cells".
3. To the pulsed cells add peptide at final concentration of 1–2 μM. Add PBS an equivalent volume to the non-pulsed target cells.
4. Incubate the cells in a 37 °C water bath for 1 h.
5. Wash the cells with RPMI medium.

6. Centrifuge the cells at $500 \times g$ for 10 min.
7. Resuspend the cell pellet in PBS and wash it two more times to assure removing the remaining peptide.
8. Resuspend the cells in PBS at a concentration of 5×10^7 cells/mL.
9. Thaw an aliquot of 5 mM stock CFSE solution.
10. Make a fresh CFSE^{low} stock solution by diluting 5 mM stock 1/10 in DMSO.
11. Incubate the unpulsed target splenocytes with the higher concentration of CFSE (CFSE^{high}): add 1 μL of the 5 mM stock CFSE for each mL of peptide-pulsed target splenocytes (final concentration of 5 μM). With a pipette resuspend cells homogeneously.
12. Incubate the peptide-pulsed splenocytes with the lower concentration of CFSE (CFSE^{low}): add 1 μL of 0.5 mM stock CFSE for each milliliter of control-pulsed cells (final concentration of 5 μM). Mix with a pipette.
13. Incubate the cells at 37 °C for 10 min. Periodically agitate the cells.
14. Add 10 volumes of pre-warmed RPMI to the CFSE-labeled cells to stop the reaction.
15. Centrifuge cells at $500 \times g$ for 5–10 min at 4 °C.
16. Remove the supernatant and resuspend the pellet in RPMI. Wash two more times with RPMI.
17. Count the mononuclear cells by trypan blue exclusion using a hemocytometer.
18. Resuspend the cells in PBS at a concentration of 5×10^7 mononuclear cells per milliliter.
19. Combine an equal volume of peptide-pulsed CFSE^{low} cells with control-pulsed CFSE^{high} cells.
20. Inject intravenously 200 μL of the combined cell populations in to the tail vein of each recipient animal.
21. After 20 h harvest spleen from the recipient mouse injected with CFSE-labeled target cells following **steps 1–10** of Subheading 3.9.
22. Resuspend the cells in FACS buffer.
23. Proceed to flow cytometry to determine the percentage of cells having low and high content of CFSE. CFSE can be analyzed using a flow cytometer equipped with 488 nm excitation and emission filters. Set the flow cytometry gates using a naïve recipient mouse (not treated with adenovirus) than has been injected with the two target populations.

24. Draw a gate identifying CFSE^{low} cells and other for the CFSE^{high} cells.
25. Collect 5,000–10,000 total CFSE positive cells.
26. Determine the percentage of CFSE^{high} and CFSE^{low} cells.
27. Calculate the ratio of unpulsed to pulsed target cells in the naïve mouse. This will be defined the 0 % lysis level.
28. The percent of specific lysis is determined by loss of the antigen-pulsed CFSE^{low} population compared with the non-pulsed CFSE^{high} control population using the next formula:

$$\left[1-\left(\text{ratio in na ve mouse} / \text{ratio in experimental mouse}\right)\right]\times 100.$$

4 Notes

1. There are several commercially available kits for Ad titration based on anti-hexon staining (e.g., Clontech X-rapid Titer; Agilent Ad-Easy Viral Titer, Cell Biolabs Quick Titer, etc.).
2. During this time cells are infected, attach, and form a monolayer. The time to achieve a maximum accumulation of virus protein will be longer when tittering in less permissive cells lines or for viruses with lower replication efficacy. If these parameters are unknown, a 48 h incubation can be tried. The differential titer of one virus in two different cell lines is a parameter that indicates the permissiveness of the cells to that particular virus.
3. Count groups of adjacent green cells as a single infectious unit as these adjacent cells derive from a single infectious unit.
4. If plaque assay is done in HEK293 cells or a cell line that does not attach well to the plate this step may be skipped to maintain the monolayer.
5. Alternatively, stain with Thiazolyl Blue Tetrazolium Bromid (MTT) adding 0.1 mL/well of a 5 mg/mL MTT solution prepared in PBS or DMEM 5 % FBS, and incubating in darkness at 37 °C and 5 % CO_2 for 3 h.
6. This assay is simplified if a vector with a reporter gene is used (e.g., GFP, luciferase) as it offers a direct read out of the amount of reporter virus (FACS, fluorescence, or luminescence).
7. A549 cells are also a good substrate cell line to detect Nab, and they attach better than HEK293 cells to the plastic. If these cells are used we recommend using and MOI of 2.5 iu/cell for the assay instead of 0.25 iu/cell. For standardization and comparison between assays a reference neutralizing polyclonal antibody (e.g., ab6982, Abcam) should be used as a test control antiserum.

8. Pre-wetting the ELISptot plate is an optional step to improve antibody binding, sensibility, and quality of spots. Nitrocellulose plates should not be treated with ethanol, only PVDF plates. MSIP PDVF plates should be treated with 15 μL of 35 % ethanol per well for 1 min. MAIPSWU PDVF plates should be treated with 50 μL of 70 % ethanol per well for 2 min.

Acknowledgments

This work was supported by a grant from the Spanish Ministry of Education and Science BIO2011-30299-C02-01, BIO2008-04692-C03-01 and from the Generalitat de Catalunya 2009SGR1212.

References

1. Choi JW, Lee JS, Kim SW, Yun CO (2012) Evolution of oncolytic adenovirus for cancer treatment. Adv Drug Deliv Rev 64(8): 720–729
2. Pesonen S, Kangasniemi L, Hemminki A (2011) Oncolytic adenoviruses for the treatment of human cancer: focus on translational and clinical data. Mol Pharm 8:12–28
3. Wang Y, Hallden G, Hill R, Anand A, Liu TC, Francis J et al (2003) E3 gene manipulations affect oncolytic adenovirus activity in immunocompetent tumor models. Nat Biotechnol 21:1328–1335
4. Funston GM, Kallioinen SE, de Felipe P, Ryan MD, Iggo RD (2008) Expression of heterologous genes in oncolytic adenoviruses using picornaviral 2A sequences that trigger ribosome skipping. J Gen Virol 89:389–396
5. Guedan S, Grases D, Rojas JJ, Gros A, Vilardell F, Vile R et al (2012) GALV expression enhances the therapeutic efficacy of an oncolytic adenovirus by inducing cell fusion and enhancing virus distribution. Gene Ther 19:1048–1057
6. Guedan S, Rojas JJ, Gros A, Mercade E, Cascallo M, Alemany R (2010) Hyaluronidase expression by an oncolytic adenovirus enhances its intratumoral spread and suppresses tumor growth. Mol Ther 18:1275–1283
7. Kast WM, Melief CJ (1991) Fine peptide specificity of cytotoxic T lymphocytes directed against adenovirus-induced tumours and peptide-MHC binding. Int J Cancer Suppl 6:90–94
8. Toes RE, Offringa R, Blom RJ, Brandt RM, van der Eb AJ, Melief CJ et al (1995) An adenovirus type 5 early region 1B-encoded CTL epitope-mediating tumor eradication by CTL clones is down-modulated by an activated ras oncogene. J Immunol 154:3396–3405

Chapter 10

The Analysis of Innate Immune Response to Adenovirus Using Antibody Arrays

Nelson C. Di Paolo and Dmitry M. Shayakhmetov

Abstract

Even though natural infections with adenovirus (Ad) are largely harmless in humans, an intravenous Ad vector administration for gene delivery purposes, especially at high doses, stimulates strong innate and adaptive immune responses, and can be fatal to the host. In animal models, intravenous Ad administration has been shown to induce transcription and release in the serum of a great number of pro-inflammatory cytokines and chemokines. Macrophages, including tissue residential macrophages (e.g., Kupffer cells in the liver), and dendritic cells throughout the body are considered to be the primary source of these pro-inflammatory mediators following their transduction with Ads. Here, we provide an overview and methodology for the qualitative and quantitative analyses of pro-inflammatory cytokine and chemokine expression in the spleen and their release into the bloodstream after intravenous Ad delivery using antibody arrays.

Key words Inflammation, Innate immunity, Adenovirus vectors, Systemic administration

1 Introduction

To date, accumulating evidence suggests that Ad triggers highly complex multifaceted innate immune and inflammatory responses, which are reflected at a clinical level in cytokinemia, thrombocytopenia, complement activation, disseminated intravascular coagulation, and multiple organ failure due to (at least in part) collateral damage from infiltrating pro-inflammatory leukocytes. Despite considerable recent progress in defining early mediators of Ad-induced inflammation [1–6], the unifying mechanistic description of sequential steps of events that lead from early Ad-host cell interactions to clinical signs of Ad-triggered systemic toxicity remains illusive.

Within minutes after intravenous administration, Ad particles are trapped by tissue residential macrophages that trigger activation and release on numerous pro-inflammatory cytokines and chemokines that orchestrate host responses targeted at recruitment of polymorphonuclear leukocytes, destruction of Ad particles, and

Miguel Chillón and Assumpció Bosch (eds.), *Adenovirus: Methods and Protocols*, Methods in Molecular Biology, vol. 1089, DOI 10.1007/978-1-62703-679-5_10,

restoration of tissue homeostasis. To this end, rapid clearance of Ad from circulation by Kupffer cells may have a protective role against the dissemination of Ads to lymphoid organs, therefore reducing systemic inflammation. In several gene therapy clinical trials, serum levels of IL-6, IL-10, and IL-1 were elevated [7–10] after systemic Ad administration at high doses (2×10^{12} to 6×10^{13} virus particles). However the role of these cytokines in the initiation of an immediate innate immune response remains unclear. A histological evaluation of tissues, including lung, liver, and spleen, revealed areas of leukocyte and neutrophil infiltration as well as infarcts ([11, 12] and our unpublished observation), indicating that most tissues in the body are involved in the inflammatory response to Ad after its intravascular administration. It has been widely accepted for a long time that Ad-mediated liver damage plays a central role in the pathogenesis of acute systemic inflammation caused by intravenous Ad administration. Specifically, it has been found that the activation of MIP-2 chemokine is critical for neutrophil attraction to liver tissue, and the inactivation of MIP-2 with an anti-MIP-2 antibody ameliorates liver pathology after intravenous Ad administration [12].

In this chapter, we focus on the analysis of inflammatory cytokine and chemokine activation using ProteomeProfiler™ antibody arrays (R&D Systems, Minneapolis, MN). We successfully used Mouse Cytokine Array panel A (Cat. No. ARY006) for simultaneous qualitative evaluation of expression for 40 different inflammatory cytokines and chemokine in mouse tissues, serum, and plasma. Here, we also describe the approach to adapt this assay for quantitative analyses of the amounts of select inflammatory cytokines and chemokines in mouse spleen after intravenous Ad administration [6, 13].

2 Materials

2.1 Standard Qualitative Cytokine and Chemokine Analysis

1. Phosphate-Buffered Saline (PBS), 20 mM NaCl, 2.68 mM KCl, 10 mM Na_2PO_4, and 1.76 mM KH_2PO_4 (pH 7.4).
2. PBS with protease inhibitors: 10 μg/mL Aprotinin, 10 μg/mL Leupeptin, and 10 μg/mL Pepstatin.
3. Triton® X-100.
4. Pipettes and sterile pipette tips.
5. Gloves.
6. Deionized water.
7. Rocking platform shaker.
8. Table-top microcentrifuge.
9. A plastic container with the capacity to hold 50 mL (for washing the arrays).

10. Plastic transparent sheet protector (trimmed to 10 × 12 cm and open on three sides).
11. Plastic wrap.
12. Absorbent lab wipes (KimWipes® or equivalent).
13. Paper towels.
14. Autoradiography cassette.
15. Film developer.
16. X-ray film (Kodak® BioMax™ Light-1, Catalog # 1788207) or equivalent.
17. Flat-tipped tweezers.
18. Flatbed scanner with transparency adapter capable of transmission mode.
19. Computer capable of running image analysis software and Microsoft® Excel.
20. Lysis buffer: PBS with inhibitors (pH 7.4), 1 % NP-40, 0.5 % sodium deoxycholate, 1 mM Na_3VO_4, 0.1 % SDS, 1 mM EDTA, 1 mM EGTA, 20 mM NaF, 1 mM PMSF.
21. ProteomeProfiler™ antibody array kit, Mouse Cytokine Array Panel A (Catalog # ARY006; R&D Systems, Minneapolis MN).
22. Tissue homogenizer (OMNI-TH, TH115 or equivalent).

2.2 Quantitative Cytokine and Chemokine Analysis

1. Phosphate-Buffered Saline (PBS), 20 mM NaCl, 2.68 mM KCl, 10 mM Na_2PO_4, and 1.76 mM KH_2PO_4 (pH 7.4).
2. Recombinant purified mouse cytokines IL-1α (Peprotech, Catalog #211-11A), IL-1β (Peprotech, Catalog #211-11B), IL-6 (Peprotech, Catalog #216-16).
3. Recombinant purified mouse chemokines MCP1 (Peprotech, Catalog # 250-10), KC (Peprotech, Catalog # 250-11), MIP2 (Peprotech, Catalog #250-15).

3 Methods

The main principle of simultaneous detection of numerous cytokines and chemokines in tissues or body fluids relies on interaction of these cytokines and chemokines with specific capture antibodies that have been spotted in duplicate on nitrocellulose membranes. Tissue homogenates, serum, or plasma samples are diluted and mixed with a cocktail of biotinylated detection antibodies. The sample/antibody mixture is then incubated with the Mouse Cytokine Array membrane supplied with ProteomeProfiler™ antibody array kit. Any cytokine/detection antibody complex present is bound by its cognate immobilized capture antibody on the membrane. Following a wash to remove unbound material, Streptavidin-HRP, and chemiluminescent detection reagents are added sequentially. The light is produced at each spot in proportion to the amount of cytokine bound and can be quantified by exposing membranes to the film with subsequent quantification of the size and density of the light-producing areas by histogram tool of imaging software or by other available techniques. The detailed description of kit components, methods, standard handling and optimization procedures are provided within product description to ProteomeProfiler™ Antibody Array Kit (R&D Systems). Here we describe our optimized methods for using this kit for the qualitative and quantitative analyses of cytokines and chemokines in mouse spleen and plasma after mouse infection with Ad.

3.1 Analysis of the Relative Amounts of Pro-Inflammatory Cytokines and Chemokines in Mouse Spleen After Intravenous Ad Injection

1. Mice are injected with Ad via a tail vein infusion and sacrificed by cervical dislocation to harvest tissues for subsequent analyses 1 h after virus administration (*see* **Note 1**).
2. Spleen is recovered, cleaned from any fat-containing surrounding tissues, and 50 mg of total splenic tissue is quickly placed into 2 mL of lysis buffer and homogenized with a tissue homogenizer.
3. Add Triton X-100 to a final concentration of 1 %. Freeze samples at ≤−70 °C, thaw, and centrifuge at 10,000 ×g for 5 min to remove cellular debris.
4. Proceed to preparation of an antibody array membrane by pipetting 2.0 mL of Array Buffer 6 into each well of the 4-Well Multi-dish supplied within the ProteomeProfiler™ kit. Array Buffer 6 serves as a block buffer.
5. Using flat-tip tweezers, remove antibody array membrane from between the protective sheets and place in a well of the 4-Well Multi-dish. The number on the membrane should be facing upward.
6. Incubate for 1 h on a rocking platform shaker. Orient the tray so that each membrane rocks end to end in its well.
7. While the membranes are blocking, prepare samples by adding up to 1 mL of each sample to 1 mL of Array Buffer 4 in separate tubes. Adjust to a final volume of 2 mL with Array Buffer 6 as necessary.
8. Add 15 μL of reconstituted Mouse Cytokine Array Panel A Detection Antibody Cocktail to each prepared sample. Mix and incubate at room temperature for 1 h.
9. Aspirate Array Buffer 6 from the wells of the 4-Well Multi-dish and add sample/antibody mixtures prepared in **steps** 7 and **8**. Cover the 4-Well Multi-dish with the lid.
10. Incubate overnight at 2–8 °C on a rocking platform shaker.

11. Carefully remove each membrane and place into individual plastic containers with 20 mL of 1× Wash Buffer. Rinse the 4-Well Multi-dish with deionized water and dry thoroughly.
12. Wash each membrane with 1× Wash Buffer for 10 min on a rocking platform shaker. Repeat two times for a total of three washes.
13. Dilute the Streptavidin-HRP in Array Buffer 6 using the dilution factor on the vial label. Pipette 2.0 mL of diluted Streptavidin-HRP into each well of the 4-Well Multi-dish.
14. Carefully remove each membrane from its wash container. Allow excess buffer to drain from the membrane. Return the membrane to the 4-Well Multi-dish containing the diluted Streptavidin-HRP. Cover the wells with the lid.
15. Incubate for 30 min at room temperature on a rocking platform shaker.
16. Wash each array as described in **steps 11** and **12**.
17. Carefully remove each membrane from its wash container. Allow excess Wash Buffer to drain from the membrane by blotting the lower edge onto paper towels. Place each membrane on the bottom sheet of the plastic sheet protector with the identification number facing up.
18. Pipette 1 mL of the prepared Chemi Reagent Mix evenly onto each membrane.
19. Carefully cover with the top sheet of the plastic sheet protector. Gently smooth out any air bubbles and ensure Chemi Reagent Mix is spread evenly to all corners of each membrane. Incubate for 1 min.
20. Position paper towels on top and sides of plastic sheet protector containing the membranes and carefully squeeze out excess Chemi Reagent Mix.
21. Remove the top plastic sheet protector and carefully lay an absorbent lab wipe on top of the membranes to blot off any remaining Chemi Reagent Mix.
22. Leaving the membranes on the bottom plastic sheet protector, cover the membranes with plastic wrap taking care to gently smooth out any air bubbles. Wrap the excess plastic wrap around the back of the sheet protector so that the membranes and sheet protector are completely wrapped.
23. Place the membranes with the identification numbers facing up in an autoradiography film cassette.
24. Expose membranes to X-ray film for 15 s to 10 min with multiple exposure times.
25. Identify the presence of specific cytokines and chemokines in the sample by superimposing the resultant membrane

chemiluminescent image on the film to a transparency overlay template, provided with ProteomeProfiler™ antibody array kit, that defines the location of capture antibody spots on the membrane for each individual cytokine and chemokine.

3.2 Analysis of the Relative Amounts of Pro-Inflammatory Cytokines and Chemokines in Mouse Plasma After Intravenous Ad Injection

1. Mice are injected with Ad via a tail vein infusion and sacrificed by cervical dislocation to collect plasma for subsequent analyses 1 h after virus administration (*see* **Note 1**).
2. Use 20 mM EDTA in PBS (without calcium and magnesium) or heparin as anticoagulant. Mix blood collected directly from the heart or from vena cava well with EDTA (2 mM final concentration) or heparin solution (10 U/mL final concentration) and place on ice while all samples are collected.
3. Centrifuge blood in a table-top microcentrifuge at 1,000 × *g* for 5 min at +4 °C.
4. Carefully collect plasma atop of the pelleted blood cells into a fresh tube. Avoid any cellular material.
5. Mix 100 mL of fresh plasma to a mixture of 1 mL of Array Buffer 4 and 0.9 mL of Array Buffer 6.
6. Follow preparation of antibody array membranes, all solutions, incubation times with buffers and all washing procedures as described in Subheading 3.1 from **steps 4–25**.

3.3 Quantitative Analysis of Specific Inflammatory Cytokines and Chemokines in Mouse Samples After Ad Infection Using Antibody Arrays

Although simultaneous detection of 40 different mouse cytokines and chemokines in mouse samples with ProteomeProfiler™ antibody arrays provides highly reliable data in a quick and cost-effective manner, the manufacturer (R&D Systems, Inc.) did not aim this method to be used for quantitative detection of pro-inflammatory mediators. We adjusted and optimized antibody array methodology with using these same reagents and kits for quantitative detection of certain cytokines and chemokines, which were prominently activated after mouse injection with Ad [6, 13]. The basis for quantitative evaluation of specific cytokines and chemokines of interest in mouse samples was the standardization of experimental conditions for incubation of the antibody array membrane with negative control tissue samples spiked with different known concentrations of purified cytokine analyte and plotting calibration curve that depicts the proportional relationship between the amount of the cytokine added to a negative control sample and the histogram signal on the film, detected after its exposure to the membrane for an exactly controlled periods of time.

1. Harvest spleen from mouse injected with PBS (instead of Ad) 1 h after PBS administration (*see* **Note 1**).
2. Mix homogenized spleen sample with serial dilutions of specific cytokines and chemokines, such as IL-1a, IL-1b, IL-6, KC, MIP-2, or MCP-1, in a range of concentrations from 10 to 10 ng.

3. Follow preparation of antibody array membranes, all solutions, incubation times with buffers, and all washing procedures as described in Subheading 3.1 from **steps 4–25**.
4. Expose membranes to the film for 15 s, 30 s, 1 min, 3 min, 5 min, and 10 min.
5. Select a set of films that were exposes to the membrane for exactly the same period of time and show the most significant difference in the dynamic range of a histogram density units, and plot a calibration curve for each particular cytokine and chemokine (*see* **Notes 2** and **3**). An example of the results produced for calibrating concentrations of MIP-2, MCP-1, and KC is shown in Fig. 1.
6. Follow procedures in an exactly the same way while processing the tissue samples, collected after Ad injection, and expose the membranes to the film for exactly the same time as calibration curve membranes.

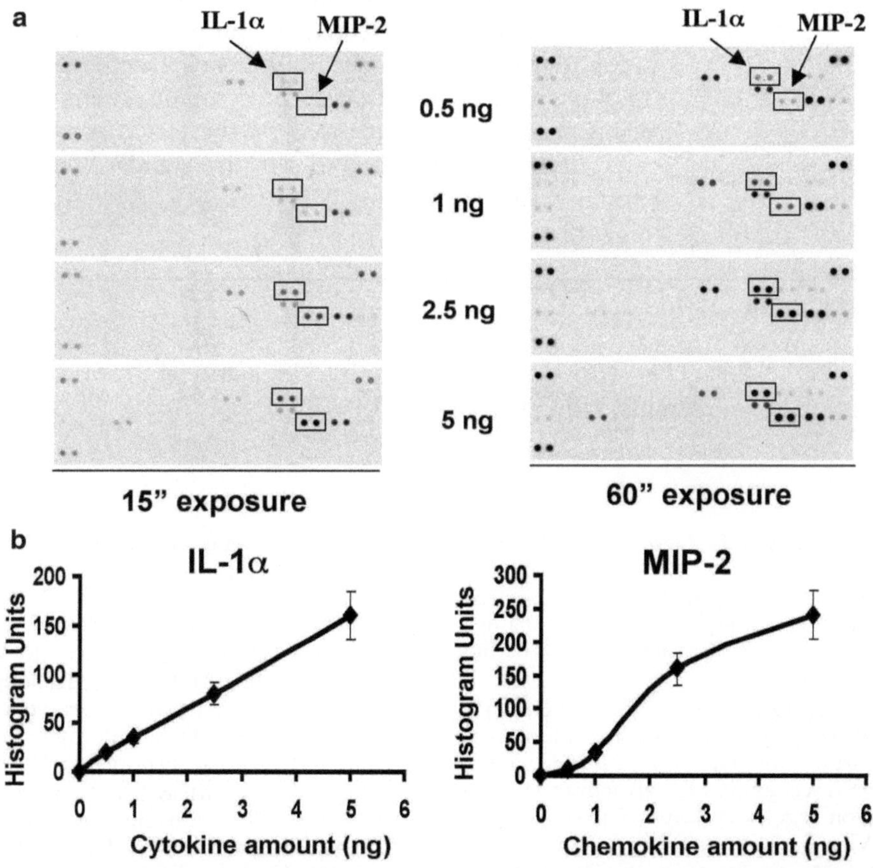

Fig. 1 Detection of increasing concentrations of recombinant purified murine IL-1α and MIP-2 chemokines, spiked with 50 mg of spleen sample, using (**a**) ProteomeProfiler™ antibody arrays, and (**b**) calibration curves derived from quantifying pixel density of indicated spots on the membrane for IL-1α and MIP-2 using histogram tool of the Adobe Photoshop software. Representative blots, exposed for 15 and 60 s to corresponding membranes are shown

7. Define the histogram density of the spots on the film for a particular cytokine in the experimental samples and calculate the amount of the cytokine present in the sample by using calibration curves from **step 5**.

4 Notes

1. All experiments involving live animals must be reviewed and approved by the institution animal care and use committee prior to conducting any such experiments, including appropriate methods for euthanasia.
2. This should be done independently for each different tissue of interest, as in each one of them tissue-specific inhibitors of protein-capture/developing antibody interaction can be present at various and unpredictable levels.
3. We have done so not only to quantify those proteins of interest which we could detect in our samples, but also for proteins that were not evidently detected in our experimental samples to prove that when a protein has not being detected (specifically TNF-α), it was truly because the cytokine/chemokine had not being produced, as evidenced by the fact that we could definitely establish a specific and sensitive standard curve for every protein tested so far.

Acknowledgments

This work was supported by US National Institutes of Health grants AI065429 and CA141439.

References

1. Fejer G, Drechsel L, Liese J, Schleicher U, Ruzsics Z, Imelli N, Greber UF, Keck S, Hildenbrand B, Krug A et al (2008) Key role of splenic myeloid DCs in the IFN-alpha beta response to adenoviruses in vivo. PLoS Pathog 4(11):e1000208
2. Nociari M, Ocheretina O, Schoggins JW, Falck-Pedersen E (2007) Sensing infection by adenovirus: toll-like receptor-independent viral DNA recognition signals activation of the interferon regulatory factor 3 master regulator. J Virol 81:4145–4157
3. Zhu JG, Huang XP, Yang YP (2007) Innate immune response to adenoviral vectors is mediated by both toll-like receptor-dependent and -independent pathways. J Virol 81:3170–3180
4. Zhu JG, Huang XP, Yang YP (2007) Type IIFN signaling on both B and CD4 T cells is required for protective antibody response to adenovirus. J Immunol 178:3505–3510
5. Muruve DA, Petrilli V, Zaiss AK, White LR, Clark SA, Ross PJ, Parks RJ, Tschopp J (2008) The inflammasome recognizes cytosolic microbial and host DNA and triggers an innate immune response. Nature 452:103–U111
6. Di Paolo NC, Miao EA, Iwakura Y, Murali-Krishna K, Aderem A, Flavell RA, Papayannopoulou T, Shayakhmetov DM (2009) Virus binding to a plasma membrane receptor triggers interleukin-1 alpha-mediated proinflammatory macrophage response in vivo. Immunity 31:110–121

7. Reid T, Galanis E, Abbruzzese J, Sze D, Andrews J, Romel L, Hatfield M, Rubin J, Kirn D (2001) Intra-arterial administration of a replication-selective adenovirus (dl1520) in patients with colorectal carcinoma metastatic to the liver: a phase I trial. Gene Ther 8:1618–1626
8. Crystal RG, Harvey BG, Wisnivesky JP, O'Donoghue KA, Chu KW, Maroni J, Muscat JC, Pippo AL, Wright CE, Kaner RJ et al (2002) Analysis of risk factors for local delivery of low- and intermediate-dose adenovirus gene transfer vectors to individuals with a spectrum of comorbid conditions. Hum Gene Ther 13:65–100
9. Ben-Gary H, McKinney RL, Rosengart T, Lesser ML, Crystal RG (2002) Systemic interleukin-6 responses following administration of adenovirus gene transfer vectors to humans by different routes. Mol Ther 6:287–297
10. Mickelson CA (2000) Department of Health and Human Services, National Institutes of Health Recombinant DNA Advisory Committee. Minutes of meeting March 8–10, 2000. Hum Gene Ther 11:2159–2192
11. McCoy RD, Davidson BL, Roessler BJ, Huffnagle GB, Janich SL, Laing TJ, Simon RH (1995) Pulmonary inflammation induced by incomplete or inactivated adenoviral particles. Hum Gene Ther 6:1553–1560
12. Muruve DA, Barnes MJ, Stillman IE, Libermann TA (1999) Adenoviral gene therapy leads to rapid induction of multiple chemokines and acute neutrophil-dependent hepatic injury in vivo. Hum Gene Ther 10:965–976
13. Doronin K, Flatt JW, Di Paolo NC, Khare R, Kalyuzhniy O, Acchione M, Sumida JP, Ohto U, Shimizu T, Akashi-Takamura S et al (2012) Coagulation factor X activates innate immunity to human species C adenovirus. Science 338:795–798

Chapter 11

Engineering Adenovirus Genome by Bacterial Artificial Chromosome (BAC) Technology

Zsolt Ruzsics, Frederic Lemnitzer, and Christian Thirion

Abstract

Bacterial artificial chromosomes (BACs) are recombinant DNA molecules designed for propagation of large and instable foreign DNA fragment in *Escherichia coli*. BACs are used in genetics of large DNA viruses such as herpes and baculoviruses for propagation and manipulation of complex genomic regions or even entire viral genomes in one piece. Viral genomes in BACs are ready for the advanced tools of *E. coli* genetics. These techniques based on homologous or site-specific recombination allow engineering of virtually any kind of genetic changes. In the recent years, BAC technology was also adapted to manipulation of adenovirus genomes and became an effective alternative to traditional genetic engineering of recombinant adenoviruses.

Key words Bacterial artificial chromosome, Recombineering, Homologous recombination in *E. coli*, Red recombinases, Flp/FRT recombination, Recombinant adenoviruses

1 Introduction

Vectors based on human adenoviruses (Ads) of species C, in particular the best-characterized serotype Ad5, appeared as extremely useful vehicles for gene therapy and vaccination. In the past few years, however, there has been growing interest in studying Ad vectors derived from species other than C. This was partially hampered by the lack of convenient systems for maintenance and manipulation of their recombinant genomes. This motivated us to develop a bacterial artificial chromosome (BAC)-based technology for Ads [1], which provides an unparalleled access to genome of large DNA viruses utilizing the advanced methodology of bacterial genetics [2].

In adenovirus genetics, BAC technology is especially useful for the establishment of reverse genetics systems when the Ad genomes under study are unstable in traditional high-copy vector systems (i.e., Ad19a, Ad3) [1, 3]. In addition, the BAC technology allows construction of recombinant genomes using a little aliquot of Ad

Miguel Chillón and Assumpció Bosch (eds.), *Adenovirus: Methods and Protocols*, Methods in Molecular Biology, vol. 1089, DOI 10.1007/978-1-62703-679-5_11, © Springer Science+Business Media, LLC 2014

DNA prepared from early passage isolates providing a fast access to molecular clones of virtually any Ad strain or variant including mutants from the past or isolates with a high risk of changes induced by adaptation to tissue culture [1, 4]. As BAC technology appeared to be very useful for fast and correct manipulation of Ad genomes, most of the applications were adapted (back) to Ad5 as well [5–7].

Here we present the basic methodology to manipulate and construct Ad-BACs by a modified recombineering, which is one of the most advanced technologies available for manipulating recombinant DNA in *E. coli* by homologous recombination. Recombineering was described as a two-step approach [8], in which homologous recombination is mediated by the red system of bacteriophage lambda and mutant selection is facilitated by the *galK* marker [8, 9]. We slightly modified this protocol combining the *galK* marker with an antibiotic selection [3, 10], which made the procedure simpler for viral BACs (*see* Fig. 1a). Since use and also construction of recombinant Ad become an everyday protocol in many molecular biology laboratories, we also included a simplified BAC-based protocol to construct first generation recombinant Ad vectors in a single step by Flp recombination. This protocol was adapted from the originally described approach for construction recombinant herpesviruses [2, 11, 12] and working for Ad5 BACs with virtually 100 % efficiency. The gene of interest is first cloned into a small donor plasmid with standard cloning techniques and this construct is unified with a BAC acceptor in *E. coli* by conditional expression of Flp recombinase via their FRT sites [11, 13].

2 Materials

2.1 Bacteriological Media for Propagation and Mutagenesis of BACs

1. Prepare LB medium, LB plates, and M9 medium according to Sambrook and Russell [14]. Add 25 μg/mL chloramphenicol (for selection of BAC plasmid), 50 μg/mL ampicillin (for selection of pCP20 plasmid), and 25 μg/mL kanamycin (for selection of inserted markers).
2. 5× M36 salts: dissolve 10 g $(NH_4)_2SO_4$, 68 g KH_2PO_4, 2.5 mg $FeSO_4 \cdot 7H_2O$ in 900 mL deionized water and adjust the pH to 7 with KOH. Add deionized water to 1,000 mL and autoclave.
3. $MgSO_4$ stock solution: prepare 1 M $MgSO_4 \cdot 7H_2O$ solution in deionized water. Sterilize by filtration.
4. D-Biotin stock solution: prepare 1 mg/mL D-biotin solution in deionized water. Sterilize by filtration.
5. L-Leucine stock solution: add L-leucine to deionized water to 10 mg/mL final concentration, heat to 50 °C until leucine is fully dissolved. Then, cool down the solution and filter sterilize.

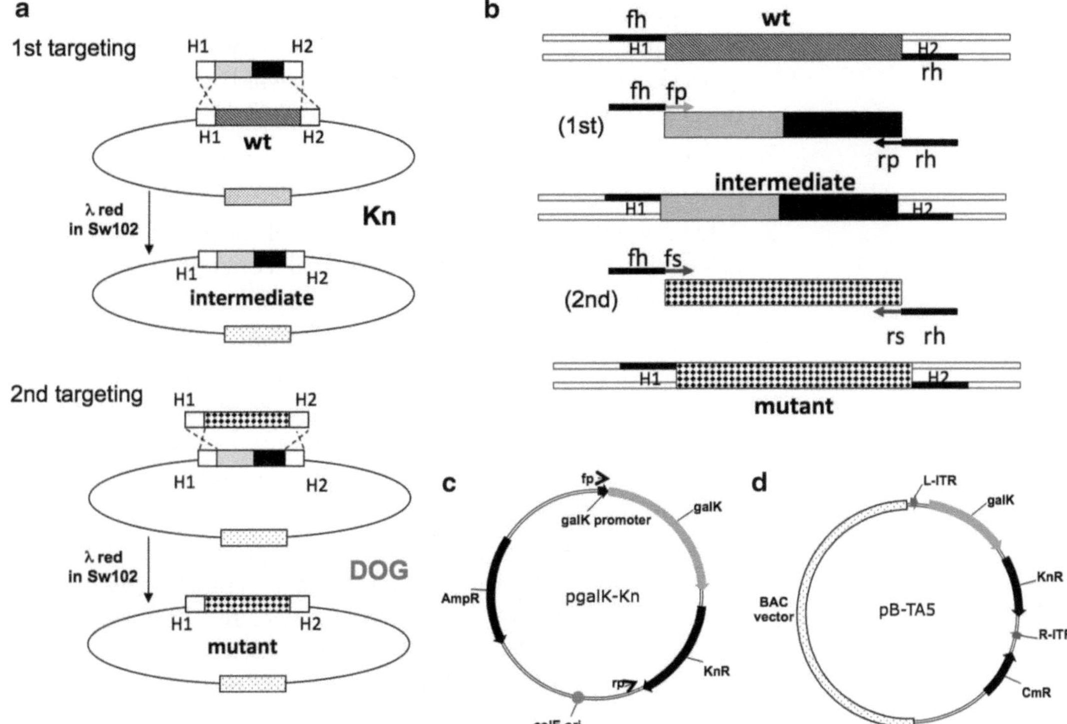

Fig. 1 (a) Schematic representation of the modified recombineering: The 1st targeting delivers a galK-Kn cassette (combined *gray* and *black boxes*) over red mediated recombination in SW102 strain to targeted BAC sequence (*hatched box*) between the selected homologies (indicated by *white boxes* and H1 and H2). For this, a PCR-derived linear fragment carrying the galK-Kn cassette flanked by the same homology regions is used. BAC vector is indicated by a *dotted box*. The resulting intermediate can be selected by kanamycin selection (Kn). The second targeting delivers a double-stranded linear DNA fragment carrying a sequence with desired mutation (*diamond patterned box*) flanked by the same homology regions (H1 and H2). This will replace the gakK-Kn cassette thereby rendering the mutant BAC Gal-, which can survive the 2-deoxy-galactose (DOG) counter selection, as it suppresses the growth of the Gal+ clones carrying the intermediate constructs. (**b**) Design of homology-flanked primers for amplification of *galK-kan* cassette: the sequence to be modified (*hatched box*, wt) is flanked by up and downstream homologies (H1 and H2, respectively). For the first targeting (1st) pgalK-Kn cassette is amplified by homology-flanked primers, which consist of two parts. In the upper primer, 50 nucleotide from the upper homology strand H1 (fh) is fused to the priming sequence, which is specific to the 5′ end of the galK-Kn cassette (fp). In the lower primer, 50 nucleotide from the complementary homology strand H2 (rh) is fused to the 5′ of the priming sequence, which is specific to the 3′ end of the galK-Kn cassette (rp). In the same way, PCR fragment for the second targeting (2nd) is amplified by homology-flanked primers, which contains the same parts. In the upper primer, 50 nucleotide from the upper homology strand H1 (fh) is fused to the priming sequence, specific to the 5′ end of the mutant sequence (diamond patterned box) (sp). In the lower primer, 50 nucleotide from the complementary homology strand H2 (rh) is fused to the 5′ of the priming sequence, which is specific to the 3′ end of the mutant sequence (rs). (**c**) Map of the marker template plasmid pgalK-Kn. Priming sites (used for the design of the homology-flanked primers in the first targeting) are indicated by *arrows* (fp and rp). (**d**) Schematic representation of the recombineering targeting intermediate BAC construct pB-TA5, which can be used to construct Ad5-BAC from genomic Ad DNA. Here the Ad left and right ITRs (L-IRT and R-ITR) serve as homologies for the replacement of the galK-Kn cassette by the genomic Ad DNA

6. 10 % glycerol: add molecular biology grade glycerol to deionized water to final concentration of 10 % (w/v) and autoclave. Cool down on ice for at least 3 h, if it will be used for preparation of competent cells (*see* Subheading 3.4).
7. Galactose stock solution: add D-galactose to deionized water to final concentration of 10 % (w/v) and autoclave.
8. 2-Deoxy-galactose (DOG) stock solution: add 2-deoxy-galactose to deionized water to final concentration of 10 % (w/v). Sterilize by filtration.
9. MacConkey plates: Add 10 g MacConkey agar (Difco) to 225 mL deionized water and autoclave. Transfer the hot agar solution into the 50 °C water bath. After reaching 50 °C, add 25 mL galactose stock solution. Mix gently and add antibiotics, if required. Mix again and poor plates using ~20 mL medium for each.
10. DOG plates Mix 4 g bacteriological grade agar (Difco) to 200 mL deionized water and autoclave. Transfer the hot agar solution into the 50 °C water bath. After reaching 50 °C, add 50 mL 5× M63 salts, 0.5 mL $MgSO_4$ stock solution, 0.25 mL D-biotin stock solution, 1.1 mL L-leucine stock solution, 5 mL 10 % glycerol, and 5 mL DOG stock solution. Mix gently and finally add antibiotics, if required. Mix again and poor plates using ~20 mL medium for each.
11. For washing competent cells use molecular biology grade deionized water, which is sterilized by autoclaving.
12. *E. coli* strain SW102 is available from Neel Copeland's laboratory (inquire through http://recombineering.ncifcrf.gov).

2.2 Molecular Biology and Cell Culture

1. All molecular biological techniques and tools needed for these protocols are standard and should be carried out according to Sambrook and Russell [14].
2. For PCR reactions, use HPLC purified primers and high fidelity polymerases.
3. BAC DNA preparation should be carried out by advanced column purification kits according to the manufacturer's instructions.
4. Dulbecco's Modified Eagle's Medium (DMEM) (Invitrogen) supplemented with 10 % fetal bovine serum (FCS) is used for culturing of 293 cells.
5. For transfection of activated Ad-BACs into 293 cells, Lipofectamine 2000 (Invitrogen) or equivalent transfection reagents should be used according to the manufacturer's instructions.

3 Methods

3.1 First Targeting to Generate the Recombineering Intermediate

1. Transform *E. coli* strain SW102 with the target Ad-BAC by electroporation (*see* **Note 1**), incubate transformants on selective LB plates, isolate single colonies and test them for integrity (*see* **Note 2**). Alternatively, plate out frozen stocks of SW102 cells already transformed with the target BAC on selective LB plates.
2. Design homology-flanked primers for amplification of *galK-kan* cassette of pgalK-Kn (*see* Fig. 1b for details of primer design and Fig. 1c for pgalK-Kn map). Insert the 50 nucleotide (nt) target sequence from the upper strand, which is directly upstream to the sequence to be modified (replaced or deleted), at the 5′ end of the forward priming site (5′-CCTGTTGACAATTAATCATCGGCA-3′). In the same way, insert the downstream 50 nt homology from the complementary strand to the 5′ end of the reverse primer (5′-GCCAGTGTTACAACCAATTAACC-3′) (*see* **Note 3**).
3. Perform PCR on 5 ng of pgalK-Kn template using the homology-flanked primers by a proofreading polymerase (*see* **Note 4**).
4. Digest the PCR product by *Dpn*I to remove template molecules from the reaction (*see* **Note 5**). Set up the digestion reaction using the entire volume of the PCR diluted at least three times (normally the final volume of the digestion reaction is 300 μL for a 100 μL PCR reaction). Add 60 U (20 U for 100 μL of final volume) of *Dpn*I and run the reaction overnight at 37 °C.
5. Purify the *Dpn*I-treated PCR product by any commercial clean-up columns and verify its integrity by agarose gel electrophoresis.
6. Inoculate 5 mL LB containing 25 μg/mL chloramphenicol with 3–4 colonies of the transformed/plated SW102 containing the desired target BAC. Incubate at 32 °C overnight in a shaking incubator.
7. Add 500 μL of overnight culture into each of two culture flasks with 25 mL LB containing 25 μg/mL chloramphenicol and continue the incubation at 32 °C in a shaking incubator. In parallel turn on the shaking water bath at 42 °C.
8. Incubate the 25 mL cultures at 32 °C until they reach $OD_{600\,nm}$ of 0.55–0.6 (takes approximately 3 h). Then, transfer the flask to the shaking water bath and heat shock the cultures at 42 °C for exactly 15 min (*see* **Note 6**).

9. Immediately after leaving the cultures in the shaking waterbath, cool down the 50 mL Falcon tube centrifuge with the rotor to +1 °C. Prepare the ice pocket.
10. Cool down the induced culture on ice immediately after the heat shock by transferring the flask directly to the ice pocket and keep them on ice for a further 15 min. Transfer each culture to a separate 50 mL Falcon tube.
11. Centrifuge the cells for 5 min at 7,000 g at 1 °C. Remove all supernatant and resuspend the bacterial pellet well in 1 mL ice-cold sterile water by pipetting up and down. Add 45 mL ice-cold sterile water.
12. Repeat **step 11** two times. The bacterial pellet will get loose, be careful to keep them in the tube while removing the supernatants after the centrifugations.
13. Centrifuge the cells for 5 min at 7,000 $\times g$ at 1 °C. Pour off all the supernatant and without inverting the tube blot out the remaining fluid by a Kleenex tissue from the upper 2/3 of the wall of the tube. Put the bacterial pellet back on ice for 1–2 min. Resuspend the bacterial pellet by pipetting up and down in the remaining supernatant and check the volume of the suspension. It should be between 120 and 150 μL (if less: add some ice-cold sterile water to balance the volume). Keep induced bacterial suspensions on ice until electroporation. Use 60–70 μL/electroporation. So, one Falcon tube contains material for two transformations and enables to generate one construct with one control reaction.
14. Transfer 60–70 μL of competent cells (from **step 13**) and ~400 ng of purified homology-flanked *galK-kan* cassette (**step 5**) to a 2 mm electroporation cuvettes and pulse the mixes with 2,500 V, 200 W, and 25 μF in Gene Pulser (or equivalent electroporation equipment). When accomplished, mix bacteria with 1 mL LB without any antibiotic and transfer them into 1.5 mL Eppendorf tubes. Incubate at 32 °C for 1 h in a shaking incubator.
15. Plate out 10 and 100 μL of the bacterial suspension on LB plates containing 25 μg/mL chloramphenicol and 25 μg/mL kanamycin. Incubate the cultures overnight at 32 °C (or over-weekend on the lab-bench).
16. Pick single colonies and streak them onto MacConkey plates containing galactose (*see* **Note** 7), 25 μg/mL chloramphenicol and 25 μg/μL kanamycin to obtain single colonies. Bright pink/red colonies (Gal$^+$) can be further processed; white colonies are either hitch-hikers with failed targeting or derivatives of recombination cassettes with mutated *galK gene*.
17. Repeat **step 16** once more (*see* **Note 8**).

18. Pick 6–12 single pink/red colonies and inoculate 10 mL LB containing 25 μg/mL chloramphenicol and 25 μg/μL kanamycin. Incubate the cultures at 32 °C overnight in a shaking incubator. Prepare BAC DNA in small scale using appropriate purification kit (*see* **Note 9**) and subject them to restriction analysis to confirm their integrity and save glycerol stocks from two verified colonies (*see* **Note 10**).

3.2 Second Targeting for Generating the Mutants

1. Synthesize the DNA sequence to be inserted into the target site either via gene synthesis service or by PCR. Sub-cloned and modified fragments can also be reintroduced into the BAC genome after amplification of the desired fragment by PCR. For this PCR also homology-flanked primers are designed in the same way as for the first targeting (*see* Fig. 1b for details). Larger fragments derived from viral DNA or even intact genomic DNA can be introduced into the appropriate recombineering intermediate (*see* **Note 11**). The linear fragments used here need at least 50 nt of homology at each end of the regions flanking the galK-Kn cassette of the recombineering intermediate. If larger DNA fragments are used, longer homologies may increase the efficiency.
2. Use two Gal + clones verified by restriction analysis from previous **step 18** (*see* **Note 12**) and inoculate 5 mL LB containing chloramphenicol and kanamycin and incubate overnight at 32 °C in a shaking incubator.
3. Repeat **steps 7–13** from Subheading 3.1 to obtain electrocompetent bacteria. However, here preparing for the second targeting, use always LB containing 25 μg/mL chloramphenicol and 25 μg/mL kanamycin for all cultures.
4. Transform 60–70 μL of the induced electrocompetent bacteria with 200 ng up to 1.5 μg of desired linear fragment (**step 1**) by an electroshock at 2,500 V, 200 W, and 25 μF in Gene Pulser (or equivalent electroporation equipment) (*see* **Note 13**). Prepare the control reaction by electroporating a mixture containing the same volume of the competent cells in sterile water and the volume of the DNA fragment. Then, mix the electroporated bacteria with 1 mL LB without any antibiotic and transfer into 50 mL Falcon tubes containing 10 mL LB without antibiotics. Outgrowth the cultures for 4.5 h at 32 °C in a shaking incubator.
5. Transfer 1 mL culture from each outgrowth to a 2 mL Eppendorf tube and spin at top speed for 15 s in a microcentrifuge at room temperature. Remove the supernatant.
6. Resuspend pellet in 1 mL M9 medium and repeat the centrifugation as in **step 5** and remove the supernatant.
7. Repeat **step 6** two times.

8. Resuspend the pellet in 1 mL M9 medium and plate out 1, 10, and 100 μL of the washed suspension from both the main and the control reactions on DOG plates containing 25 μg/mL chloramphenicol (*see* **Note 14**). Incubate the plates at 32 °C for 2–4 days.
9. Pick 10–20 single colonies and streak each one onto a quarter of an LB plate containing 25 μL chloramphenicol. Incubate the plates at 32 °C overnight.
10. Pick single colonies from each quarter to inoculate 10 mL LB containing 25 μg/mL chloramphenicol and incubate the cultures overnight at 32 °C.
11. Prepare BAC DNA by the small-scale protocol (*see* **Note 9**) using 9.8 mL of the overnight culture and subject them to a restriction pattern analysis. Save the remaining BAC cultures at +4 °C.
12. Streak a loop of the remaining cultures corresponding to two colonies with correct restriction pattern onto MacConkey plates (*see* **Note 7**) supplemented with galactose and 25 μg/mL chloramphenicol.
13. Pick a well-isolated white colony from each plate and propagate the BACs for sequencing and virus reconstitution.

3.3 Construction of Recombinant Adenoviruses by Flp Recombination in *E. coli*

1. In this protocol the gene of interest is cloned into a shuttle plasmid, which is unified with a modified Ad-BAC by Flp recombination in *E. coli* via their FRT sites (*see* Figs. 2a, b). Therefore, the gene of interest, or the ORF to be expressed, needs to be cloned first into the vector pO6A5-CMV (*see* **Note 15** and Fig. 2c).
2. Retransform the FRT ready Ad-BAC vector (pBA5-FRT, *see* **Note 16**) to DH10B strain of *E. coli* (*see* **Note 17**) or plate out a small aliquot of frozen stock of the FRT-containing Ad-BAC vector on LB plates containing 25 μg/mL chloramphenicol.
3. Prepare electrocompetent cells from a single colony of DH10B cultures of the pBA5-FRT (*see* **Note 18**).
4. Transform the competent DH10B cells carrying the pBA5-FRT with 5 ng of pCP20 [13] (*see* **Note 19**) and plate out 10 μL the 1 mL outgrowth cultures on LB plates containing 25 μL chloramphenicol and 50 μg/mL ampicillin and incubate them overnight at 32 °C. Alternatively, plate out a small aliquot of frozen stock of the DH10B cells carrying both the pBA5-FRT and pCP20 on LB plate containing 25 μg/mL chloramphenicol and 50 μg/mL ampicillin and incubate them at 32 °C overnight (*see* **Note 20**).
5. Inoculate 5 mL LB containing 25 μg/mL chloramphenicol and 50 μg/mL ampicillin with a single colony from the plates

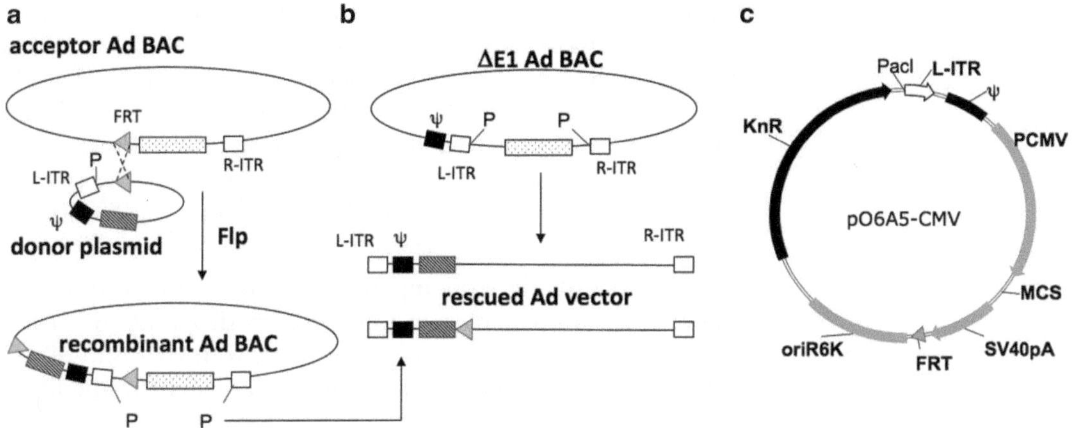

Fig. 2 (a) Schematic representation of the Flp/FRT recombination for construction of first generation Ad vectors. This approach is based on Flp-mediated unification (recombination of two components: An acceptor Ad-BAC and a donor plasmid. The acceptor Ad-BAC carries the full Ad genome except its very left terminus including the left ITR (L-IRT) the packaging signal (Ψ) and the E1 region (not shown). Instead, in this construct the left terminal sequences are replaced by a FRT site. The donor plasmid which possesses a complementary FRT site carries the left terminal Ad sequences sufficient to reconstitute a first generation Ad vector (L-ITR and packaging signal) is fused to a transcription unit for the gene of interest (*hatched box*). After Flp recombination the donor plasmid is entirely inserted into the acceptor BAC. (**b**) Activation of a traditionally designed E1 deleted Ad-BAC vector: the linear Ad sequence is released by digestion of *PacI* restriction endonuclease at ITR adjacent *PacI* sites (P). A similar Ad genome can be released from the recombinant Ad-BACs generated by the Flp approach after the same treatment. The only difference is a single FRT site between the transgene and the rest of the vector genome. Other donor plasmid sequences are left behind connected to the BAC vector sequences (*dotted box* in part A). (**c**) Map of the cloning vector p06A5-CMV used for construction of the donor plasmids for gene expression. The gene of interest can be inserted into the multiple cloning site (MCS)

with the DH10B cells carrying pBA5-FRT and pCP20. Incubate the culture overnight at 32 °C.

6. Add 1 mL of overnight culture into two culture flasks with 200 mL LB containing 25 μg/mL chloramphenicol and 50 μg/mL ampicillin and continue the incubation at 32 °C in a shaking incubator until they reach density of $OD_{600\ nm}$ of 0.55–0.6 (takes approximately 3 h).
7. Cool the induced culture down immediately by transferring the flask directly to the ice pocket and keep them on ice for further 15 min. Transfer the culture to four 50 mL Falcon tube.
8. Centrifuge the cells for 5 min at 7,000 × *g* on +1 °C. Remove all supernatant and resuspend the bacterial pellet well in 1 mL ice-cold sterile 10 % glycerol by pipetting up and down. Collect the four suspensions in one 50 mL Falcon tube and add 45 mL ice-cold sterile 10 % glycerol.
9. Repeat **step 8** twice. The bacterial pellet will get loose, be careful to keep them in the tube while removing the supernatants after centrifugations.

10. Centrifuge the cells for 5 min at 7,000×*g* on +1 °C. Pour off all the supernatant and without inverting the tube blot out the remaining fluid by a Kleenex tissue from the upper 2/3 of the wall of the tube. Put the bacterial pellet back on ice for 1–2 min. Resuspend the bacterial pellet by pipetting up and down in the remaining supernatant and check the volume of the suspension. It should be in between 120 and 150 μL. Add ~600 μL of ice-cold sterile 10 % glycerol to get a final volume of 700 μL.
11. Aliquot a 35 μL electrocompetent cell suspension into 0.5 mL Eppendorf tubes and snap-freeze them dropping the tubes directly into liquid nitrogen. Store the frozen vials on −80 °C. Alternatively, keep on ice the aliquots, that will be immediately used.
12. For each expression construct thaw up or reserve one vial of pBA5-FRTxpCP20 electrocompetent cells on ice together with one vial for the control.
13. Transfer one aliquot of the electrocompetent cells and ~100 ng of the pO6A5-CMV construct (**step 1**) to 0.2 mm electroporation cuvette on ice and mix carefully. Add sterile deionized water instead of the donor plasmid into the control cuvette.
14. Electroporate mixes with 2,500 V, 200 W, and 25 μF in Gene Pulser (or equivalent electroporation equipment). When accomplished, mix bacteria with 1 mL LB without any antibiotic and transfer them into 1.5 mL Eppendorf tubes. Incubate at 43 °C for 1 h in a shaking incubator.
15. Plate out 10 and 100 μL of the bacterial suspension on LB plates containing 25 μg/mL chloramphenicol and 25 μg/mL kanamycin. Incubate the cultures overnight at 43 °C (*see* **Note 20**).
16. Inoculate 2–3 flasks with 100 mL LB medium supplied with 25 μg/mL chloramphenicol and 25 μg/mL kanamycin with a single colony from the plate incubated at 43 °C and grow culture overnight at 37 °C with shaking (*see* **Note 21**).
17. Prepare BAC DNA using an appropriate purification kit according to the manufacturer's instructions (*see* **Note 9**). Verify the constructs by restriction analysis and sequencing.

3.4 Reconstitution of Recombinant Adenoviruses from BACs

1. Prepare 10–30 μg BAC DNA from the desired recombinant using an appropriate purification kit according to the manufacturer's instructions (*see* **Note 9**). It is recommended to rescue two independent verified clones for the same construct to ensure consistent data in upstream experiments.
2. For virus reconstitution activate 10 μg of Ad-BAC DNA by linearization with the appropriate restriction endonuclease, cutting Ad-BACs at specific ITR-flanking sites. Usually this

activation is made by the *Pac*I restriction endonuclease (for most Ad5-derived constructs including the here described pBA5-FRT-derived constructs, *see* Fig. 2b) in a 100 μL standard reaction.

3. Add 50 μL phenol/chloroform and vortex for 20 s.
4. Spin tubes in a microcentrifuge at maximum speed for 5 min at room temperature.
5. Transfer 90 μL of the aqueous upper phase into a fresh tube.
6. Add 10 μL 3 M Na-acetate (pH 4.5) and 200 μL 100 % ethanol. Mix with flicking by finger tips. The precipitated DNA should become visible immediately. Incubate for 5 min at room temperature.
7. Spin down tubes in a microcentrifuge at maximum speed for 10 min at room temperature.
8. Remove supernatant completely and immediately dissolve pellet in 20 μL sterile deionized water.
9. The day before transfection plate 2.5×10^5/well 293 cells to a 6-well plate (*see* **Note 22**).
10. Transfect 293 cells with 6–8 μL (ca. 2–3 μg) activated Ad-BAC DNA using any commercial transfection reagent (optimized for 293 cells) according to the manufacturer's instructions. Incubate the transfected plates for 3 days under standard cell culture conditions.
11. On the second day after transfection plate out 2.5×10^5/well 293 cells to a 6-well plate (*see* **Note 22**).
12. On the third day after transfection (*see* **Note 23**) harvest cells by rinsing the plate with the medium several times until all cells are detached from the plate surface. Usually there is no need to use cell scraper, but make sure that you remove all cells from the well.
13. Collect cell suspension in a sterile 15 mL Falcon tube by centrifugation for 5 min at $200 \times g$.
14. Remove supernatant and resuspend cell pellet in 400 μL DMEM + 10 % FCS.
15. Place tube into liquid nitrogen for 3 min.
16. Immediately place tube in 37 °C water bath until cells have thawed completely.
17. Repeat **steps 15** and **16** twice.
18. Centrifuge sample for 10 min at $3{,}500 \times g$ at room temperature and transfer supernatant (lysate) into a fresh 1.5 mL Eppendorf tube.
19. Infect one or two wells of the plate from **step 11** with 200 μL of lysate for each well. Incubate plates for 3–4 days under

standard cell culture conditions until most cells show cytopathic effect (CPE) (*see* **Note 23**).

20. Repeat **steps 12–17**.
21. Centrifuge sample for 10 min at 3,500 × *g* at room temperature and transfer the new lysate into a fresh Eppendorf tube. Store the final lysate at −80 °C or directly continue with Ad amplification according to standard protocols.

4 Notes

1. To propagate electrocompetent *E. coli* SW102, plate out frozen stocks of SW102 cells on LB plates and incubate them at 32 °C overnight and carry out the **steps 5–11** of Subheading 3.3 using LB without antibiotics for all cultures. Electroporate competent cells with 10 ng of purified Ad-BAC using 2,500 V, 200 W, and 25 μF in Gene Pulser (or equivalent electroporation equipment).
2. Ad-BACs are very stable and can be easily retransformed with high fidelity. Therefore a simple restriction analysis is sufficient to ensure their integrity in a new *E. coli* strain after electrotransformation. However, we do not recommend using any other method for transformation of BACs.
3. The pgalK-Kn construct and its nucleotide sequence are available from the corresponding author upon request.
4. We recommend the Expand High Fidelity PCR System from Roche. In this case, mix the template with the 1 μL of each primers (from a 100 μM stock), 2 μL 10 mM dNTPs, 10 μL 10× PCR reaction buffer, 2.5 U polymerase, add deionized PCR quality water to final volume of 100 μL. Run the PCR for 35 cycles; in the first 18 cycles touch-down the annealing temperature from 62 to 45 °C and set the annealing temperature at 45 °C for additional 17 cycles. Allow 2 min for annealing time. Elongation in all 35 cycles should be set at 68 °C for 2 min, and denaturation in all 35 cycles should be at 94 °C for 30 s.
5. Residual PCR template is the main source of false-positive background colonies during the first targeting. Therefore, a complete *Dpn*I digestion of template plasmid after PCR is essential. Since the PCR-derived DNA is not methylated, the residual plasmid-derived template DNA can be removed by *Dpn*I digestion, which only cuts *dam* methylated DNA. To make sure that the template pgalK-Kn is well methylated, it should be propagated in *dam*-positive *E. coli* strain (DH10B).

6. The heat shock at 42 °C will induce expression of the lambda phage derived red recombinases. It is crucial to keep this induction time as rigorously as possible for reproducible results.
7. Make sure that MacConkey plates are prepared using premixed agar powders, which do not contain any sugar and thus, added galactose will be the only carbon source in the plate. We recommend using MacConkey agar from Difco.
8. It is crucial to produce pure stocks of recombineering intermediate construct for the second targeting. Do not save time avoiding this replating step(s). A few germs carrying the original target BAC contaminating the stocks of the intermediates can corrupt the second targeting!
9. As BACs are propagated in *E. coli* in single copy, typical yields of BAC DNA are usually only between 10 and 30 μg from a 100 mL overnight culture. We recommend using Nucleobond PC-20 and PC-100 (Macherey and Nagel) for small and middle scale preparations, respectively. The small-scale Nucleobond PC-20 column yields preparations sufficient for 2–3 restriction analysis and retransformation. Sequencing, detailed restriction analysis, or virus reconstitution requires larger scale Nucleobond PC-100 purification. The culture volume here can be scaled up to 200 mL if multiple assays are planned.
10. It is very valuable to have a pure, well-characterized recombineering intermediate construct. The work above can be saved for the next mutation within the same region if the intermediate construct is kept as a glycerol stock.
11. To clone the genomic adenovirus DNA as BAC, viral DNA can be purified either from the infected cells or purified virions [1, 3]. For this recombineering, a special targeting BAC construct is required, which needs to be constructed by traditional cloning procedures. This intermediate construct should contain the galK-Kn cassette flanked by the left and the right ends (few 100 nt each) of the Ad genome of interest [1, 3]. See the targeting construct pB-TA5 for cloning wild-type Ad5 genomes and/or its variants in Fig. 1d as an example. The pB-TA5 is available upon request from the corresponding author.
12. One can inoculate overnight culture from a glycerol stock using fitting intermediates from the previous experiment. In this case the recombination fragment needs to be prepared to fit to the previous intermediate.
13. As a rule of thumb, more DNA is needed if larger fragments are introduced. To start with 500 ng insert for a standard recombineering normally delivers good results. However, DNA amounts may be limited by the toxicity after transformation or by arcing during the electroshock. In these cases the DNA amounts need to be systematically reduced.

14. If a verified intermediate is used from previous experiments, it is enough to plate out 100 μL of the water control. A few colonies can be found in the control plate since the counter selection is never 100 % efficient. However, if there is not at least fivefold difference in favor to the main reaction, contamination of the intermediate stock must be considered.

15. pO6A5-CMV is a small plasmid vector with a basic CMV promoter-based transcription unit for expression of the cloned genes in mammalian cells either after transfection of the plasmid itself or transferring it into the Ad vector. The left ITR and the packaging signal of the Ad5 (in front of the CMV promoter) will provide the left end of the rescued viral genome after recombination with the BAC acceptor. It also contains the R6Kgamma origin of replication [15], which is dependent on the presence of the *pir* locus and therefore cannot be maintained in normal laboratory *E. coli* strains. *E. coli* strain PIR1, which expresses the *pi* protein encoded by the *pir* gene in trans, is required for cloning the gene of interest in pO6A5-CMV and propagation of its recombinants. This plasmid carries a 34 bp FRT site for integration (5′-GAAGTTCCTATTCTCTAGAAAGTATAGGAACTTC-3′). The PIR1 strain is available from Invitrogen and the pO6A5-CMV from Sirion Biotech.

16. When constructing normal first generation vectors by Flp recombination, the Ad genome is usually deleted from the left end to the end of the E1B gene in the BAC acceptor. Instead, a 48 bp at the FRT site (5′-GAAGTTCCTATTCCGAAGTTC CTATTCTCTAGAAAGTATAGGAACTTC-3′) is inserted, allowing insertion of donor plasmids carrying 34 bp FRT sites (*see* **Note 15**). In this design the donor will provide the leftmost region of the vector genome with ITR and the packaging signal after recombination. Thus the BAC acceptor cannot be rescued to virus. This setting can be found in the pBA5-FRT exemplified here, which is available from Sirion Biotech. The FRT site can be inserted virtually into any place of the vector genome by recombineering (Subheadings 3.1 and 3.2) in order to customize the expression locus. In this case, however, the pO6A5-CMV should also be modified, for example sequences corresponding to the ITR and the packaging signal should be removed from the donor plasmid used for internal insertions.

17. To propagate electrocompetent *E. coli* DH10B plate out frozen stocks of DH10B (Invitrogen) cells on LB plates and incubate them overnight at 37 °C and carry out the **steps 5–11** of Subheading 3.3 using LB without antibiotics at 37 °C for all cultures. Alternatively custom electrocompetent DH10B preparations can also be used. Electroporate the competent

cells by 10 ng of purified Ad-BAC using 2,500 V, 200 W, and 25 μF in Gene Pulser (or equivalent electroporation equipment).

18. To propagate electrocompetent *E. coli* DH10B carrying pBA5-FRT plate out frozen stocks on LB plates containing 25 μg/mL chloramphenicol and incubate them at 37 °C overnight or pick a colony from the plated retransformation (**step 2**) and carry out the **steps 5–11** of Subheading 3.3 using LB containing 25 μg/mL chloramphenicol at 37 °C for all cultures.
19. The pCP20 Flp expression plasmid is available from the corresponding author upon request. Do not co-transform the Ad-BAC with the pCP20 because the pCP20 also contains a chloramphenicol resistance marker. Always introduce the Ad-BAC first and overtransform this intermediate with pCP20.
20. pCP20 is maintained in *E. coli* by a temperature sensitive origin of replication, which is only active at lower culture temperatures (30–33 °C) and expresses the Flp recombinase under the control of the temperature sensitive lambda *cI857* repressor, which is induced at higher temperature (42–43 °C). This design provides a very useful conditional Flp expression system, which can be maintained in suppressed state (30–33 °C) and as soon as it is induced (42–43 °C) the plasmid is also automatically cleared from *E. coli* [13]. Alternatively, custom electrocompetent Flp ready pBA5-FRT preparations (Sirion Biotech) can also be used. In this case proceed with **step 13**.
21. Insertion of donor plasmid to the BAC acceptor by this Flp/FTR system is very efficient and results in virtually 100 % positive clones. Sometimes even two copies of the donor plasmid are inserted into the FRT site of the BAC acceptor. If the donor plasmid is inserted into the genome end (like in the pO6A5-CMV/pBA5-FRT approach described here) this does not harm the rescue because the extra copy is simply cut out by *Pac*I activation of the Ad-BAC for rescue (*see* Subheading 3.4 and Fig. 2a). However, if internal acceptor FRT sites are targeted, care should be taken to sort out double insertions. This requires an appropriate restriction analysis of 6–12 clones, which can be carried out after small-scale BAC DNA preparations. If two single insertions are identified, the procedure can continue with the **step 16** of Subheading 3.3 as described.
22. 293 cells as well as other E1 complementing cells can be used for reconstitution of BACs derived from species C and some serotypes from other species [1, 4–6]. 293 cells are robust and we recommend using them because endotoxin load may be higher than usual after BAC DNA preparations propagation of high copy plasmids. Special complementing cell line may be required for reconstitution of some BAC-derived Ad from other species [3].

23. These reconstitution times are given for Ad5-based constructs. Reconstitution of Ads from other species may need longer rescue times (5–6 days each step).

Acknowledgments

The authors would like to thank Sigrid Seelmeir and Simone Boos for their excellent technical support during establishment and development of these methodologies. We also would like to thank Harald Wodrich and Ruben Martinez (University of Bordeaux) for their generous input in testing and improvement of these techniques.

References

1. Ruzsics Z, Wagner M, Osterlehner A, Cook J, Koszinowski U, Burgert HG (2006) Transposon-assisted cloning and traceless mutagenesis of adenoviruses: development of a novel vector based on species D. J Virol 80:8100–8113
2. Ruzsics Z, Koszinowski UH (2008) Mutagenesis of the cytomegalovirus genome. Curr Top Microbiol Immunol 325:41–61
3. Sirena D, Ruzsics Z, Schaffner W, Greber UF, Hemmi S (2005) The nucleotide sequence and a first generation gene transfer vector of species B human adenovirus serotype 3. Virology 343:283–298
4. Imelli N, Ruzsics Z, Puntener D, Gastaldelli M, Greber UF (2009) Genetic reconstitution of the human adenovirus type 2 temperature-sensitive 1 mutant defective in endosomal escape. Virol J 6:174
5. Puntener D, Engelke MF, Ruzsics Z, Strunze S, Wilhelm C, Greber UF (2011) Stepwise loss of fluorescent core protein V from human adenovirus during entry into cells. J Virol 85:481–496
6. Wodrich H, Henaff D, Jammart B, Segura-Morales C, Seelmeir S, Coux O et al (2010) A capsid-encoded PPxY-motif facilitates adenovirus entry. PLoS Pathog 6:e1000808
7. Jager L, Hausl MA, Rauschhuber C, Wolf NM, Kay MA, Ehrhardt A (2009) A rapid protocol for construction and production of high-capacity adenoviral vectors. Nat Protoc 4:547–564
8. Warming S, Costantino N, Court DL, Jenkins NA, Copeland NG (2005) Simple and highly efficient BAC recombineering using galK selection. Nucleic Acids Res 33:e36
9. Alper MD, Ames BN (1975) Positive selection of mutants with deletions of the gal-chl region of the Salmonella chromosome as a screening procedure for mutagens that cause deletions. J Bacteriol 121:259–266
10. Lemnitzer F, Raschbichler V, Kolodziejczak D, Israel L, Imhof A, Bailer SM et al (2013) Mouse cytomegalovirus egress protein pM50 interacts with cellular endophilin-A2. Cell Microbiol 15:335–351
11. Bubeck A, Wagner M, Ruzsics Z, Lotzerich M, Iglesias M, Singh IR et al (2004) Comprehensive mutational analysis of a herpesvirus gene in the viral genome context reveals a region essential for virus replication. J Virol 78:8026–8035
12. Wagner M, Koszinowski UH (2004) Mutagenesis of viral BACs with linear PCR fragments (ET recombination). Methods Mol Biol 256:257–268
13. Cherepanov PP, Wackernagel W (1995) Gene disruption in Escherichia coli: TcR and KmR cassettes with the option of Flp-catalyzed excision of the antibiotic-resistance determinant. Gene 158:9–14
14. Sambrook J, Russell DW (2001) Molecular cloning: a laboratory manual, 3rd edn. Colb Spring Harbor LAboratory Press, Cold Spring Harbor, New York
15. Stalker DM, Kolter R, Helinski DR (1979) Nucleotide sequence of the region of an origin of replication of the antibiotic resistance plasmid R6K. Proc Natl Acad Sci U S A 76:1150–1154

Chapter 12

Construction, Production, and Purification of Recombinant Adenovirus Vectors

Susana Miravet, Maria Ontiveros, Jose Piedra, Cristina Penalva, Mercè Monfar, and Miguel Chillón

Abstract

Recombinant adenoviruses provide a versatile system for gene expression studies and therapeutic applications. In this chapter, a standard procedure for their generation and small-scale production is described. Homologous recombination in *E. coli* between shuttle plasmids and full-length adenovirus backbones (E1-deleted) is used for the generation of recombinant adenoviral vectors genomes. The adenovirus genomes are then analyzed to confirm their identity and integrity, and further linearized and transfected to generate a recombinant adenoviral vector in permissive human cells. These vectors are then purified by two sequential CsCl gradient centrifugations and subjected to a chromatography step in order to eliminate the CsCl and exchange buffers. Finally, the viral stock is characterized through the quantification of its viral particle content and its infectivity.

Key words Adenoviral vector, Homologous recombination, Adenovirus construction, Adenoviral vector production, Adenovirus purification

1 Introduction

Adenoviruses are non-enveloped, double-stranded DNA viruses. They have been studied in depth and very well characterized, which paved the way to the generation of recombinant adenoviruses, the most commonly used gene transfer systems with applications in many scientific areas. Modification of the adenoviral genome is limiting by use of classical molecular biology techniques. Initially, recombinant adenoviruses were generated by direct ligation of the gene of interest into the adenoviral genome. However, direct ligation was technically difficult due to the large size of the adenovirus genome (36 kb), the lack of unique restriction sites for cloning, and the low efficiency of large DNA fragment ligations. Next generation viral vectors were based on two-plasmid rescue system, where the gene of interest was first cloned in a shuttle vector containing part of the adenovirus genome, and then transferred

Miguel Chillón and Assumpció Bosch (eds.), *Adenovirus: Methods and Protocols*, Methods in Molecular Biology, vol. 1089, DOI 10.1007/978-1-62703-679-5_12,

into the vector genome by homologous recombination within an adenovirus packaging cell line [1]. Newly generated recombinants were selected by screening individual plaques in permissive packaging cells [2]. However, this strategy needed to be improved due to the low frequency of the recombination event and the need for purification of individual viral plaques, which means it was both labor-intensive and time-consuming. Homologous recombination provides a third, highly flexible to manipulate large DNA, and has found widespread use. It exploits the highly efficient homologous recombination machinery in bacteria to generate a recombinant adenovirus vector by homologous recombination in *E. coli* between a large plasmid containing most of the adenovirus genome and a small shuttle plasmid containing the expression cassette flanked by sequences homologous to the region to be targeted in the viral genome. The recombinant adenovirus genome is then linearized by restriction digestion and used to transfect E1-complementing mammalian cells to produce viral particles and propagate the vector.

From the variety of known adenoviruses, researchers have concluded that serotypes 2 (HAdV-C2) and 5 (HAdV-C5) of species C are the most effective for creating viral vectors for use in gene therapy [3]. However, the following methods can also be used to generate vectors derived from other adenovirus serotypes [4], chimeric vectors, which contain viral proteins from different serotypes [5], vectors other than first generation, such as oncolytic vectors [6], or helper and helper-dependent adenovirus vectors [7] or even, to generate nonhuman recombinant adenoviruses [8]. In all cases, the recombination procedures either in BJ5183 bacteria or yeast can be applied directly, though for each particular vector the researcher must use specific plasmids and/or specific permissive cell lines.

Cloning the gene of interest within the adenovirus genome by homologous recombination and further amplification in permissive HEK293 cells may lead to rearrangements and instability of the viral genome. It is highly recommended to analyze the recombinant adenovirus genome digesting with a large set of restriction enzymes before proceeding to amplification in HEK293 cells. Also, the recombinant adenoviral vectors amplified in HEK293 may be tested for the presence of *Mycoplasma* as a quality control measure prior to the start of the purification. At a functional level, tests detecting newly produced viral proteins such as the antibody anti-hexon staining method should be used to assess the infectivity of the vectors. In order to quantitate adenovirus yield, viral particles should be determined by a direct physicochemical method such as UV spectroscopy. The ratio between adenoviral particles and infectious units is highly relevant to assess the quality of the viral stock (ratios of ≤30 particles to 1 infectious unit are considered appropriate, though the ratios obtained with the procedures described here are usually ≤20:1).

2 Materials

2.1 Adenovirus Construction by Homologous Recombination in Bacteria

1. LB Broth: 25 g Miller's LB in 1 L of ddH_2O. Sterilize by autoclave.
2. LB + Ampicillin: Add 100 mg ampicillin to 1 L LB Broth.
3. LB + Ampicillin plates: Add 15 g agar to 1 L LB Broth. Autoclave. Cool down to 50 °C and add 100 mg ampicillin. Pour on plates.
4. *E. coli* strain BJ5183 (endA, *sbcB*$^-$, *recBC*$^-$, *str*^{R}).
5. *E. coli* strains DH5α or similar.
6. Commercial DNA kit for purifying DNA fragments from agarose gels.
7. 1 % (wt/vol) agarose gels.
8. Appropriate restriction enzymes.
9. Commercial DNA minipreparation kit.
10. Spectrophotometer.

2.2 Generation of Recombinant Adenovirus

1. Dulbecco's Modified Eagle's Medium (DMEM) supplemented with 10 or 2 % fetal bovine serum (FBS).
2. *Pac*I restriction enzyme.
3. TE buffer (10 mM Tris–HCl, 1 mM EDTA).
4. HEK-293 cells or other adenovirus packaging cell lines.
5. 10 mM polyethylenimine (PEI) 25,000, branched (Sigma, ref. 408727).
6. 150 mM NaCl.

2.3 Purification of Recombinant Adenovirus by Banding on CsCl

1. Ultracentrifuge: Beckman Coulter Optima L90 K or L100XP and SW32Ti and SW40Ti rotors (Beckman Coulter). Ultra-Clear centrifuge tubes for SW32Ti rotor (Beckman, ref. 344058) and polyallomer centrifuge tubes for SW40Ti rotor (Beckman Coulter ref. 331374).
2. CsCl solutions of densities: 1.4 g/mL, 1.34 g/mL, and 1.25 g/mL in TD buffer (137 mM NaCl, 5.1 mM KCl, 0.7 mM $Na_2HPO_4{\cdot}7H_2O$, 25 mM Tris Base, pH = 7.4 (adjusted with HCl).
3. 18 G needles, 10 mL syringes, pipette-aid, 10 mL pipettes.
4. Disposable Sephadex G-25 columns (PD-10, GE Healthcare, ref. 17-0851-01 or equivalent).
5. 1× PBS Ca^{++}/Mg^{++} (PAA ref. H15-001).
6. Glycerol anhydride.

2.4 Titration of Recombinant Adenovirus Using Anti-Ad/Hexon Staining

1. Dulbecco's Modified Eagle's Medium (DMEM) supplemented with 2 % fetal bovine serum (FBS) (PAA, ref. A15-151).
2. Primary Antibody anti-hexon 2Hx-2 from ATCC or similar antibodies.
3. FITC or Alexa488-conjugated secondary antibody.
4. Fluorescence microscope.

2.5 Quantification of Adenovirus Particles by Spectrophotometry

1. Lysis Solution: 0.1 % SDS, 1 mM EDTA in 10 mM Tris–Cl, pH 7.4.
2. Heat block.
3. Spectrophotometer and cuvettes.

3 Methods

Cloning a gene of interest by homologous recombination in bacteria is a two-step procedure. In the first step, the gene of interest is cloned into a shuttle plasmid vector using standard molecular biology methods. The shuttle plasmid contains two fragments of adenovirus genomic sequence (usually 4–5 kb from the 5′ end) flanking the multicloning site. The presence and orientation of the gene of interest is confirmed by restriction digestion analysis and/ or sequence analysis. The second step consists in introducing the gene of interest into the adenovirus genome by homologous recombination between the shuttle plasmid and a large backbone plasmid. This backbone plasmid contains most of the adenovirus genome, but lacks essential genes for virus propagation, usually, E1 genes. Rapid detection of positive recombinants is achieved by antibiotic selection and restriction digestion analysis.

3.1 Adenovirus Construction by Homologous Recombination in Bacteria

In this protocol, the recombination between the shuttle plasmid and the adenovirus genome contained in the backbone plasmid is performed in the *E. coli* strain BJ5183. Positive recombinants are selected by resistance to an antibiotic (*see* Fig. 1).

3.1.1 Homologous Recombination in Bacteria

1. Linearize the backbone plasmid (i.e., pKP1.4 [7, 9] or similar) with a restriction enzyme cutting in the insertion site, such as *Swa*I. Digestion should be made in two steps: first, digest 1.5 μg plasmid with 10 U of *Swa*I for 24 h. Then, add 10 U more of *Swa*I and digest during seven additional hours (*see* **Note 1**).
2. Check background by transforming BJ5183 bacteria with 100 ng of digested plasmid. If the number of colonies obtained is higher than five, repeat **steps 1** and **2**. After verification, store in 200 μL aliquots.

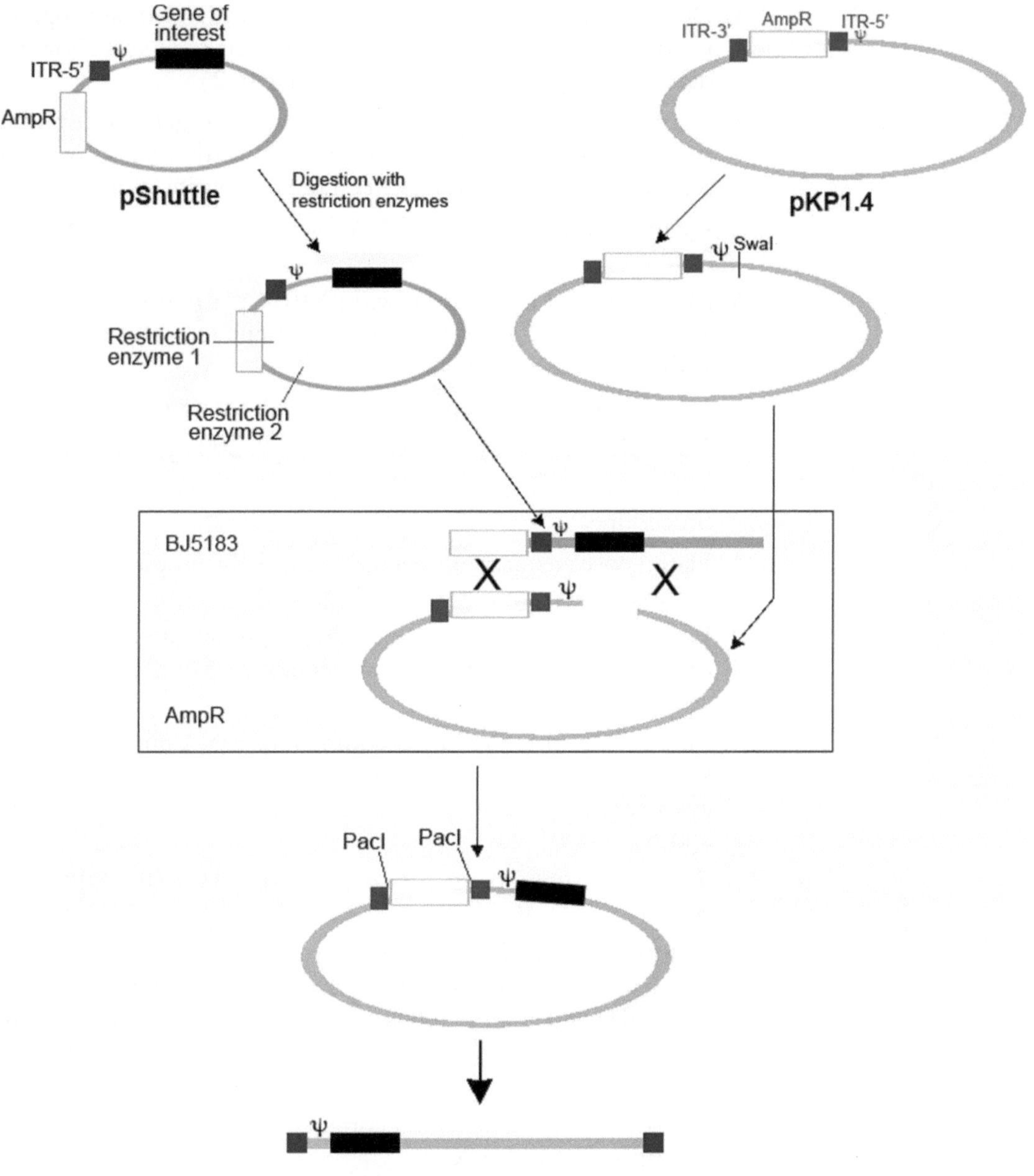

Fig. 1 Schematic outline of the homologous recombination in *E. coli* (procedure I). The shuttle plasmid already contains the gene of interest, which is flanked by the 5′ ITR and packaging signal (Ψ) in one end, and adenoviral sequences in the other. The pKP1.4 backbone plasmid contains the adenovirus genome (except the E1 region). First, the pKP1.4 plasmid is linearized by *SwaI*, and the shuttle plasmid is digested by one (RE-I) or two restriction enzymes (RE-II) in the Amp^R gene. Both digested plasmids are co-transfected in BJ5183 bacteria for homologous recombination and only bacteria carrying recombinant plasmids containing the adenoviral genome plus the gene of interest are viable in LB+Amp^R plates. For production of the viral pre-stock, recombinant plasmids are digested with *PacI* to liberate the vector genome

3. Digest 3 μg of the shuttle plasmid with two appropriate restriction enzymes. Use 10 U of each enzyme and allow the digestion to proceed overnight (*see* **Note 2**).
4. Confirm complete digestion by 1 % agarose gel electrophoresis. Purify the DNA fragment containing the expression cassette using a commercial DNA fragment purification kit.
5. Quantify the recovered DNA by measuring absorbance at 260 nm.
6. Mix gently 50 ng of linearized pKP1.4 plasmid with different amounts of the purified DNA fragment (*see* **Note 3**). Start with the following molar ratios:

   1:5 pKP:fragment (approx. 50 ng pKP: 50 ng of fragment) or

   1:20 pKP:fragment (approx. 50 ng pKP: 200 ng of fragment).
7. Transform competent BJ5183 *E. coli* bacteria, using standard procedures.
8. Plate co-transformed bacteria in LB + Ampicillin dishes. Incubate overnight at 37 °C.
9. Pick at least 12 isolated small colonies. Inoculate each in 3 mL of LB + Ampicillin. Incubate overnight at 37 °C with shaking at 220 rpm (*see* **Note 4**).
10. Purify plasmid DNA with a commercial DNA minipreparation kit. Resuspend DNA in 30 μl of Milli-Q H_2O. Check by agarose gel electrophoresis (1 %). Store at −20 °C (*see* **Notes 5** and **6**).
11. Transform competent *E. coli* (strain TOP10, DH5α or similar).
12. Culture co-transformed bacteria in LB + Ampicillin dishes. Incubate overnight at 37 °C. Pick two colonies from each plate presenting less than 1,000 colonies and inoculate each in 3 mL LB + Ampicillin. Grow overnight at 37 °C with shaking at 220 rpm.
13. Purify plasmid DNA with a commercial DNA minipreparation kit and store at −20 °C.
14. Proceed to check and identify positive recombinants with restriction enzymes (Subheading 3.1.2).

3.1.2 Identity of the Vector Genome by Restriction Enzyme Analysis

Integrity and identity of the vector genome can be quickly analyzed by restriction enzyme digestion. Since each gene of interest has a specific DNA sequence, a set of informative restriction enzymes must be previously chosen by comparing with a sequence analysis program the expected recombinant adenovirus to the original backbone plasmid.

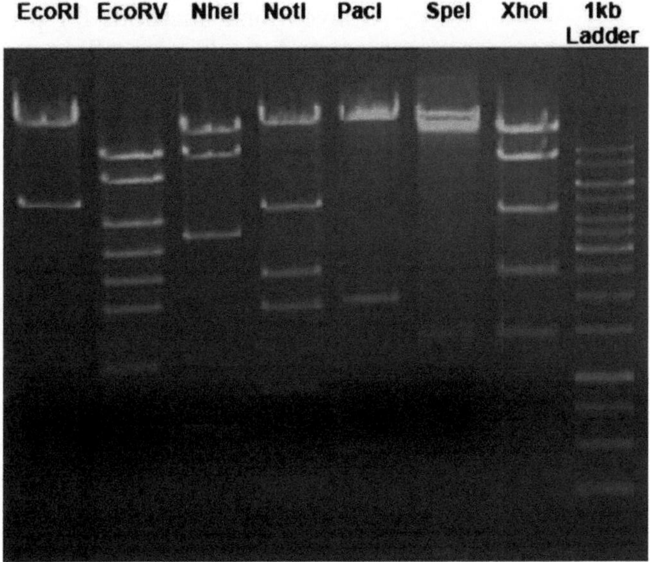

Fig. 2 Analysis of the integrity and identity of the vector genome by multiple restriction enzyme digestion. Informative restriction enzymes must be previously chosen with a Sequence Analysis computer program

1. Digest 1 μg of purified plasmid DNA from the colonies selected after homologous recombination with 0.3–0.5 U of an informative restriction enzyme, for 4 h.
2. Perform digestions with at least seven or eight different enzymes and run in a 1 % agarose gel (*see* Fig. 2).
3. If one restriction enzyme pattern does not correspond with the expected pattern, repeat the digestion. If the observed pattern still does not correspond with the expected pattern or if there are more than one unexpected enzyme patterns, discard the selected DNA.
4. Inoculate one positive recombinant in 200 mL of LB + Ampicillin. Purify plasmid DNA with a commercial DNA maxipreparation kit and store at −20 °C.

3.2 Generation and Amplification of Recombinant Adenovirus

Recombinant adenovirus vectors are replication-defective and, therefore, they must be produced and propagated in cell lines that complement the defect. Even so, these vectors must be handled following Biosafety Level 2 practices, just like the wild-type adenoviruses from which they are derived. The reason is that replication-competent adenoviruses (RCA) can result from rare recombination events between the recombinant vector genome and the adenoviral sequences present in HEK293 cells during the amplification process. Appropriate information and guidance can be obtained from each institution's Biosafety Committee and/or the Occupational & Environmental Safety Office.

Conventional methods for producing small volumes of viral vectors involve culturing cells in stationary, adherent cultures. The main protocols of this small-scale production method are presented below, while methods for large-scale production of Ad viral vectors can be found in the literature. Recombinant adenoviruses obtained at the end of the protocol are ready to be used in experimental procedures involving cells or laboratory animals, as well as a starting material for subsequent rounds of larger amplifications.

1. Twenty-four hours before transfection, plate 10^6 HEK-293 cells/well in a six-well cell culture plate with DMEM + 10 % FBS (Fetal Bovine Serum). Three wells are needed for each adenovirus vector, one of them as control.
2. Digest 100 μg of recombinant adenoviral plasmid with 15 U of *Pac*I for 5 h to separate the adenovirus genome from the remaining plasmid sequences.
3. Precipitate digested DNA with 0.1 volume of sodium acetate and 2.5 volumes of ethanol and resuspend in 100 μL sterile TE buffer.
4. Perform a standard transfection using 6 μg of *Pac*I digested plasmid per 10^6 HEK293 cells (*see* **Note 7**).
5. Incubate at 37 °C and 5 % CO_2 for 3 days (*see* **Note 8**).
6. Scrape/harvest medium and cells. Freeze at –80 °C and thaw at 37 °C three times to lyse the cells and release the adenoviruses.
7. Centrifuge at room temperature for 5 min and 1,200 × *g* and keep the supernatant (crude lysate). Discard the pellet.
8. Use all the obtained crude lysate to infect 7×10^6 HEK293 cells (at 70–80 % confluency) in a 10-cm plate, in a final volume of 8 mL, with DMEM + 2 % FBS.
9. Incubate at 37 °C and 5 % CO_2 until general cytopathic effect is observed (usually between 4 and 9 days).
10. Harvest medium and cells. Freeze–thaw three times to release adenovirus from cells.
11. Centrifuge at room temperature for 5 min and 1,200 × *g*. Discard the pellet and keep the supernatant (crude lysate).
12. Use 1 mL of the crude lysate obtained in **step 11** to infect a total 3.5×10^8 HEK293 cells in twenty 15-cm plates (70–80 % confluency), in a final volume of 14 mL/plate with DMEM + 2 % FBS.
13. Incubate at 37 °C and 5 % CO_2 until general cytopathic effect is observed (*see* **Note 9**).

14. Harvest medium and cells. Distribute it among six 50-mL Falcon tubes. Centrifuge at room temperature for 5 min at 620 × *g*. Keep 19 mL of the supernatant and discard the rest.
15. Resuspend the pellets with the reserved 19 mL of supernatant.
16. Freeze/thaw three times to release adenovirus from cells.
17. Centrifuge at room temperature for 5 min at 1,200 × *g*. Discard the pellet and keep the supernatant (crude lysate).
18. This crude lysate can be tested for *Mycoplasma* contamination as a quality control (*see* **Note 10**).
19. Store at −80 °C.

3.3 Purification of Recombinant Adenovirus

Protocols for purification of adenovirus vectors have evolved over the last decade. The most classical and easy method for a non-specialized laboratory remains the ultracentrifugation on a CsCl gradient, although this purification method is limited by the volume of cell lysate that can be processed. However, this procedure is still widely used and most of the time is sufficient for fundamental studies and/or early in vivo preclinical evaluation of the vectors. More complex techniques based on column chromatography and membrane techniques are now well developed for the generation of high purity grade and up-scaled production suitable for human clinical applications [10, 11].

3.3.1 Initial Step Gradient

1. In a 38.5 mL Ultra-clear SW32 polyallomer centrifuge tube, add 10 mL of 1.25 g/mL of CsCl.
2. Add 10 mL of 1.4 g/mL of CsCl by placing the tip of a 10 mL pipette at the bottom of the tube (and under the first CsCl solution), then carefully and slowly dispensing solution to create two phases.
3. Gently add ~19 mL of crude lysate on top of the 1.25 g/mL CsCl layer. Leave about 0.1 cm at the top of the tube. If a counterbalance tube is required, use the same total volume of D-PBS.
4. Balance tubes by weighing and load in rotor.
5. Centrifuge for 1 h 42 min at 125,500 × *g* 18 °C in a Beckman SW32 rotor, maximum brake.
6. Remove tubes from rotor with forceps.
7. The adenovirus appears as an opaque band at the interface of the 1.25 and the 1.4 g/mL CsCl layers (*see* Fig. 3a). Remove the band by piercing the tube about 1 cm below the vector with a 10 mL syringe with a 18 G needle.
8. Collect all the vector bands in a 15-mL Falcon tube and add 1.34 g/mL CsCl to obtain a final volume of 13 mL.

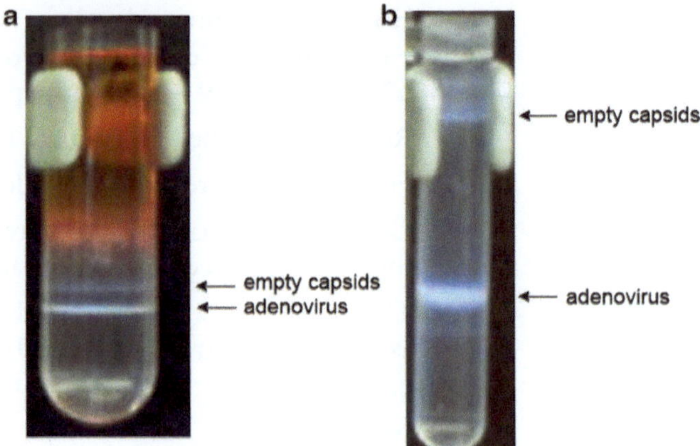

Fig. 3 Visualization of adenovirus and empty capsids after the initial step gradient (**a**) and after the second isopycnic gradient (**b**), respectively

3.3.2 Second Isopycnic Gradient

1. Transfer the recovered vector band from the previous step into a new polyallomer centrifuge tube for SW40 Ti rotor.
2. In a second tube add CsCl (1.34 g/mL) to balance the rotor. Balance the tubes and load in rotor.
3. Centrifuge for 22 h at 155,000 × *g* 18 °C in a Beckman SW40 rotor, maximum brake.
4. Remove tubes from rotor with forceps.
5. The vector appears as an opaque band near the center of the tube (*see* Fig. 3b). Remove band as described above in a maximum of 2.5 mL.

3.3.3 Desalting Column and Storage

1. Prepare a PD-10 column following the manufacturer's instructions. Load up to 2.5 mL of purified adenovirus on the column.
2. Label 1.5 mL tubes. Start to collect eluted fractions by adding 0.5 mL of PBS 1× Ca^{2+}/Mg^{2+}. Repeat this seven times. The adenovirus is clearly visible as an opaque solution between fractions 4 and 7.
3. Add sterile glycerol to the fractions to a final concentration of 10 %.
4. Titrate fractions 3–8 using anti-Ad/hexon (*see* Subheading 3.4.1).
5. Select and pool the desired fractions.
6. Aliquot in small volumes in 0.5 mL tubes and store at −80 °C as quickly as possible (*see* **Note 11**).

3.4 Adenovirus Characterization

3.4.1 Titration of Recombinant Adenovirus Using Anti-Ad/Hexon Staining

The most common infectious assay involves the detection of a cytotoxic effect on cells following viral spread. The fastest infectious assay relies on the transfer of a specific gene coding for a protein and the subsequent detection of this protein. Other infectious assays have relied on molecular DNA detection techniques following viral DNA replication. Here we describe a titration method based on the detection of hexon viral protein, a specific gene transferred by the adenoviral vector and expressed during its infection cycle. This method can be applied to RCA quantification (*see* **Note 12**).

1. The day before, seed cells in a collagen-coated 96-well plate, in a quantity enough to obtain a confluence of 80 % on the day of the experiment (35,000 cells/well).
2. Prepare serial dilutions of the adenoviral vector in 24-well plates using DMEM + 2 % FBS (*see* **Note 13**).
3. Aspirate the medium of the cells (96-well plate prepared in **step 1**) and add 100 μL/well of each viral dilution (*see* **Note 14**).
4. Incubate virus and cells for 48 h (*see* **Note 15**).
5. Remove medium from wells very carefully to avoid cell loss. Air dry for 5–10 min.
6. Add 75 μL/well of ice-cold methanol. Incubate for 10 min at –20 °C.
7. Aspirate methanol. Wash cells twice with 100 μL/well of D-PBS + 1 % BSA.
8. Add 50 μL/well of primary Antibody diluted in D-PBS + 1 % BSA (*see* **Note 16**), avoiding bubble formation. Incubate for 1–2 h at 37 °C.
9. Wash twice with 100 μL/well of D-PBS + 1 % BSA.
10. Add 50 mL/well of FITC or Alexa488-conjugated secondary antibody diluted in D-PBS + 1 % BSA, avoiding bubble formation (1/300 dilution is suitable for most commercial antibodies). Using Alexa488 can increase test sensitivity. Incubate for 1–2 h at 37 °C in the dark.
11. Wash twice with 100 μL/well of D-PBS + 1 % BSA.
12. Count green cells using an inverted fluorescence microscope. A cloud of positive cells ("comet effect") is suggestive of secondary infections and, consequently, it should be counted as a single positive cell. The mean of different dilutions will be used to calculate the infectious titer. Note that only 100 μL of each dilution is being used for analysis, so a tenfold factor should be applied to obtain infectious units per milliliter (IU/mL).

3.4.2 Quantification of Adenovirus Particles by Spectrophotometry

1. Dilute samples in lysis buffer (usually 1:10 or 1:20). A control dilution with PBS Ca^{2+}/Mg^{2+} with 10 % glycerol should be prepared and used as a blank for the optical density measure.

2. Incubate the blank and the adenovirus samples for 10 min at 56 °C.
3. Centrifuge 1 min at 16,000 × *g*.
4. Measure the optical density at 260 nm.
5. The concentration of adenoviral particles is determined by multiplying the absorbance by the appropriate dilution factor and then dividing by the extinction coefficient of adenovirus (ε260 Ad5 = 9.09×10^{-13} OD mL cm $virion^{-1}$), calculated from the data of Maizel et al. [12].

4 Notes

1. This digestion should be as complete as possible in order to remove the background from undigested plasmids in the following steps.
2. The enzymes of choice must not cut within the expression cassette. As the aim of this step is to facilitate recombination, it is recommended to leave the expression cassette by at least of 1 kb of DNA sequences on both sides. Additionally, it is better if at least one enzyme cuts within the antibiotic resistance gene.
3. Use as controls (a) pKP1.4 linearized with *Swa*I, and (b) gel-purified fragment from the shuttle plasmid. It is recommended to use highly competent bacteria since BJ5183 exhibit lower transformation efficiencies than conventional *E. coli* strains.
4. Two populations of colonies are expected: large and small size colonies. Large colonies are generally background from the shuttle plasmid, while small colonies will likely contain recombinant plasmids, which are low copy number plasmids. The number of small colonies must be at least three times higher than in control plate (digested pKP1.4 only) to continue the protocol. If the number of small colonies is less than three times, start again the procedure, and also use a different ratio in **step 5**.
5. Do not store the BJ5183 bacteria after overnight growth, as unwanted recombinants might appear. Perform plasmid purification early in the morning.
6. Agarose gel electrophoresis allows the identification of colonies containing the shuttle plasmids, which will be discarded. Select only clones with high molecular weight DNA, as well as those clones with no detectable DNA, since the yield of recombinant DNA is much lower than that from background or unwanted rearrangements.
7. Common methods to transfect HEK293 cells include calcium phosphate precipitation [13] or polyethylenimine (PEI)-mediated DNA delivery [8] though other methods based on

cationic molecules can also be used. For reasons of efficiency, simplicity, and cost, PEI transfection is highly recommended. In the case that PEI is used, the following protocol for the preparation of the PEI/DNA complex can be applied:

(a) Prepare PEI and/DNA complexes in 2-mL Eppendorf tubes:

(a.1) In a tube labelled A: put 6 μg DNA and 150 μL of sterile 150 mM NaCl. Mix well.

(a.2) In a tube labelled B: put 13.5 μL PEI 10 mM and 150 μL of sterile 150 mM NaCl. Mix well.

(b) Slowly add solution B dropwise to solution A.

(c) Incubate the DNA with the transfection reagent for 20 min at room temperature.

(d) Remove growth medium from HEK-293 cells and wash once with serum-free DMEM gently. Remove DMEM and add 0.5 mL of serum-free DMEM per well.

(e) Add PEI/DNA complexes dropwise to cells. Incubate at 37 °C and 5 % CO_2 for 6 h.

(f) Remove medium and add 3 mL of fresh DMEM + 10 % FBS.

8. If the recombinant viral genome carries a fluorescent marker gene, check initial transfection efficiency as well as adenoviral infection during amplification. It is not expected to observe a visible cytopathic effect (CPE) in this step.

9. The adenovirus life cycle spans 36 h. In our conditions, CPE begins at around 30 h postinfection, peaking at approximately 40 h, which is when we usually harvest the virus. Some factors, including the producer cell line and the gene of interest expressed from the adenoviral genome, may change the CPE appearance pattern, resulting in variability in harvest times.

10. Inoculate 7.5 μl of lysate in A549 nonpermissive human cells and culture in the absence of antibiotics for several days (usually 7 days), to amplify the possible mycoplasma contamination and increase sensitivity. Then, Venor™ GeM Mycoplasma Detection Kit can be used to detect *Mycoplasma* following the instructions of the manufacturer.

11. Repeated freeze-thaw cycles should be avoided as they will cause a decrease in viral infectivity. Aliquot in small volumes to minimize this effect.

12. HEK293 cells are normally used for infectious unit quantification. The detection of RCAs requires the use of noncomplementary human cells such as A549, unlike replication-defective adenoviruses, RCAs will be able to replicate in A549 cells.

13. The range of dilutions will be selected according to the estimated concentration of the viral stock to titer. A standard scheme of serial dilutions is drawn in Table 1.

Table 1
Standard scheme of the medium (DMEM + 2 % FBS) and virus volumes for the serial dilutions of viral stock titration

	24-Well plate													
Dil #:	**1**	**2**	**3**	**4**	**5**	**6**	**7**	**8**	**9**	**10**	**11**	**12**	**13**	**14**
Medium (ml)	2	2	1.8	0.5	0.5	0.5	0.5	0.5	0.5	0.5	0.5	0.5	0.5	0.5
+	2 μl virus	2 μl #1	0.2 ml #2	0.5 ml #3	0.5 ml #4	0.5 ml #5	0.5 ml #6	0.5 ml #7	0.5 ml #8	0.5 ml #9	0.5 ml #10	0.5 ml #11	0.5 ml #12	0.5 ml #13
Relative dilution factor	1/1,000	1/1,000	1/10	1/2	1/2	1/2	1/2	1/2	1/2	1/2	1/2	1/2	1/2	1/2
Absolute dilution factor	$1/10^3$	$1/10^6$	$1/10^7$	$1/2 \cdot 10^7$	$1/4 \cdot 10^7$	$1/8 \cdot 10^7$	$1/1.6 \cdot 10^8$	$1/3.2 \cdot 10^8$	$1/6.4 \cdot 10^8$	$1/1.3 \cdot 10^9$	$1/2.6 \cdot 10^9$	$1/5.2 \cdot 10^9$	$1/10^{10}$	$1/2 \cdot 10^{10}$

14. When infecting, start adding the most diluted point and use the same tip while infecting with the same virus. Try to release the 100 μL of infection medium/well without disturbing the monolayer (particularly for HEK293 cells). Try not to make bubbles.
15. Time must be chosen depending on rate of replication of the adenovirus in the cell line used in order to avoid secondary infections.
16. For most hybridomas 1/5 dilution from supernatant is recommended. If using a purified anti-adenovirus or anti-hexon Ab, 1/500 dilution can be initially tested.

References

1. McGrory WJ, Bautista DS, Graham FL (1988) A simple technique for the rescue of early region I mutations into infectious human adenovirus type 5. Virology 163(2):614–617
2. Graham FL, Prevec L (1995) Methods for construction of adenovirus vectors. Mol Biotechnol 3:207–220
3. Kovesdi I, Brough DE, Bruder JT, Wickham TJ (1997) Adenoviral vectors for gene transfer. Curr Opin Biotechnol 8:583–589
4. Barratt-Boyes SM, Soloff AC, Gao W, Nwanegbo E, Liu X, Rajakumar PA et al (2006) Broad cellular immunity with robust memory responses to simian immunodeficiency virus following serial vaccination with adenovirus 5- and 35-based vectors. J Gen Virol 87: 139–149
5. Glasgow JN, Kremer EJ, Hemminki A, Siegal GP, Douglas JT, Curiel DT (2004) An adenovirus vector with a chimeric fiber derived from canine adenovirus type 2 displays novel tropism. Virology 324:103–116
6. Cascallo M, Alonso MM, Rojas JJ, Perez-Gimenez A, Fueyo J, Alemany R (2007) Systemic toxicity-efficacy profile of ICOVIR-5, a potent and selective oncolytic adenovirus based on the pRB pathway. Mol Ther 15:1607–1615
7. Alba R, Hearing P, Bosch A, Chillon M (2007) Differential amplification of adenovirus vectors by flanking the packaging signal with attB/attP-PhiC31 sequences: implications for helper-dependent adenovirus production. Virology 367:51–58
8. Kremer EJ, Boutin S, Chillon M, Danos O (2000) Canine adenovirus vectors: an alternative for adenovirus-mediated gene transfer. J Virol 74:505–512
9. Delenda C, Chillon M, Douar A.M, Merten OW (2007) Cells for gene therapy and vector production. In: Ralf Poertner (ed) Methods in biotechnology: animal cell biotechnology: methods and protocols. Totowa, New Jersey, USA. Humana 24:23–91
10. Burova E, Ioffe E (2005) Chromatographic purification of recombinant adenoviral and adeno-associated viral vectors: methods and implications. Gene Ther 12(Suppl 1):S5–S17
11. Silva AC, Peixoto C, Lucas T, Küppers C, Cruz PE, Alves PM, Kochanek S (2010) Adenovirus vector production and purification. Curr Gene Ther 10(6):437–455
12. Maizel JV Jr, White DO, Scharff MD (1968) The polypeptides of adenovirus. I. Evidence for multiple protein components in the virion and a comparison of types 2, 7A, and 12. Virology 36:115–125
13. Umana P, Gerdes CA, Stone D, Davis JR, Ward D, Castro MG et al (2001) Efficient FLPe recombinase enables scalable production of helper-dependent adenoviral vectors with negligible helper-virus contamination. Nat Biotechnol 19:582–585

Chapter 13

Scalable Production of Adenovirus Vectors

Ana Carina Silva, Paulo Fernandes, Marcos F.Q. Sousa, and Paula M. Alves

Abstract

Recombinant adenoviruses (AdV) are highly efficient at gene transfer for a broad spectrum of cell types and species. They became one of the vectors of choice for gene delivery and expression of foreign proteins in gene therapy and vaccination purposes. To meet the need of significant amounts of adenoviral vectors for preclinical and possibly clinical uses, scalable and reproducible production processes are required.

In this chapter, we review processes used for scalable production of two types of first generation (E1-deleted) adenoviral vectors (Human and Canine) using stirred tank bioreactors. The production of adenovirus vectors using either suspension (HEK 293) or anchorage-dependent cells (MDCK-E1) are described to exemplify scalable production processes with different cell-culture types. The downstream processes will be covered in the next chapter.

Key words Adenovirus, Human, Canine, Suspension, Microcarriers, Gene therapy, Vaccination, Stirred tank bioreactor

1 Introduction

1.1 Adenoviral Vectors

Viral vectors are currently the most efficient tools for in vivo gene transfer. Human adenovirus vectors (HAdV) derived from serotype 5 are the most characterized viruses among those of the same family [1, 2]. HAdV has been considered a good candidate for human gene therapy due to a number of advantages including its wide cell tropism in quiescent and non-quiescent cells, its inability to integrate the host genome and its high production titer [3]. Currently, adenovirus vectors (AdV) are being tested as subunit vaccine systems for numerous infectious agents ranging from malaria to HIV-1 [4, 5]. Additionally, they are being explored as vaccines against a multitude of tumor-associated antigens. However, gene transfer efficacy and the clinical use of HAdV can be hampered by the preexisting humoral and cellular immunity in most humans [6, 7]. Therefore, alternatives able to bypass some of these clinical disadvantages, while keeping the numerous advantages associated

Miguel Chillón and Assumpció Bosch (eds.), *Adenovirus: Methods and Protocols*, Methods in Molecular Biology, vol. 1089, DOI 10.1007/978-1-62703-679-5_13,

with HAdVs vectors are also under evaluation. One of these approaches is the use of nonhuman adenovirus vectors, being Canine adenovirus type 2 (CAV-2) probably the best described [6–9]. The paucity of neutralizing antibodies and memory T cells in humans and efficient gene delivery obtained for CAV-2 vectors in the central nervous system, make these viruses promising tools for human gene therapy [6, 9]. The extraordinary ability to preferentially transduce neurons combined with a remarkable capacity of axonal transport, make CAV-2 vectors candidates for the treatment of neurodegenerative diseases.

1.2 Production Process

Most of HadV production processes reported use HEK 293 cells as the cell line of choice. However, the risk of recombinant competent adenovirus (RCA) generation in those cells led to the development of other alternatives, such as PER.C6 or CAP cells, that overcome this problem by not having any overlapping sequences between E1 (expressed by cell line) and viral vector genome [10, 11]. In order to facilitate culture manipulation and process scale-up, cells grown in suspension are preferred and maximal viable cell densities, up to $3–9 \times 10^6$ cells/mL for HEK 293 and PER.C6 are routinely achieved in bioreactor cultures [12–15]. For anchorage-dependent cell lines, like MDCK-E1 used for CAV-2 production, it is required the use of microcarrier technology for production in scalable stirred tanks. The most common microcarriers used for cell growth and virus production are Cytodex™-1, but other types are available and it selection depends on the cell line and culture conditions used [16, 17]. Stirred tank bioreactors (STB) allow for a fine control of cell environment (nutrients, pH, O_2, and temperature) permitting different modes of operation (batch, fed-batch, and perfusion). The majority of HAdV production processes reported in the literature are performed using suspension cells cultured in bioreactor using batch mode with culture medium exchanged at the time of infection. Other operation modes have been explored for AdV production at high cell densities in order overcome the "cell density effect" phenomenon related with limitation in key nutrient depletion or inhibitor byproducts accumulation (*see* Fig. 1). The infection of the cells, by a well-characterized viral seed, is normally performed with cell concentrations at infection (CCI) of $1–3 \times 10^6$ cells/mL and a multiplicity of infection (MOI) of 10–200 total virus particles/cell. AdV infections are rapid and the harvest is normally done between 36 and 72 h post infection (hpi). At the end of the process it is usually achieved an amplification ratio higher than 200 with titers of AdV ranging from 10^4 to 10^5 total particles per cell which corresponds normally to $10^3–10^4$ infectious particles/cell (for detailed information see reviews [3, 18, 19]).

This chapter describes methodologies for the scalable production of two types of first generation (E1-deleted) adenoviral vectors using STBs. The production of HAdV using HEK 293 cells grown

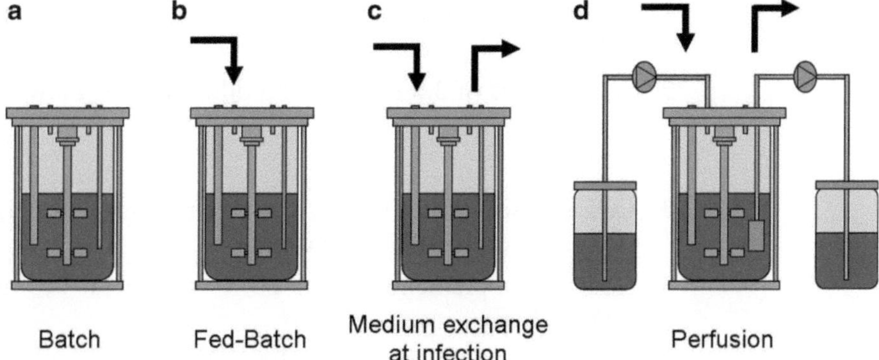

Fig. 1 Different modes of culture operation used in adenovirus production. (**a**) Batch mode: easiest approach as no extra feeding is required and the risk of contamination is low; (**b**) Fed-batch: primarily used to extend culture lifetime by supplementing limiting nutrients and/or reducing the accumulation of toxic metabolites; (**c**) Medium exchange at infection: ensures removal of toxic metabolites supplying fresh nutrients at the time of infection; (**d**) Perfusion mode: cells are retained inside the bioreactor at a relatively high cell concentration, while fresh medium is continuously supplied, allowing replenishing of nutrients and removal of toxic metabolites

in suspension and CAV-2 using MDCK-E1 cells in microcarriers are described to exemplify productions of AdV with cell lines adapted to suspension and anchorage-dependent, respectively.

2 Materials

2.1 Cell Culture

2.1.1 HEK 293 Cells Culture

1. HEK 293 cells (ATCC: CRL-1573) adapted to suspension growth (*see* **Note 1**).
2. Ex-Cell® 293 serum-free medium (Sigma-Aldrich Co. LLC) supplemented with 4 mM of glutamine (Gibco) (*see* **Note 2**).
3. 500 mL Erlenmeyer flasks (Corning Life Sciences or equivalent).
4. Orbital shaker (IKA model HS 260 or equivalent).

2.1.2 MDCK-E1 Cells Culture

1. MDCK-E1 cell line [20] derived from MDCK (ECACC, Nr 84121903) (*see* **Note 1**).
2. Optipro™ SFM supplemented with 4 mM of glutamine (both from Gibco) (*see* **Note 2**).
3. Trypsin–EDTA 0.25 %.
4. Dulbecco's phosphate-buffered saline without calcium and magnesium (D-PBS).

2.2 Monitoring and Characterization of Cells and AdV

2.2.1 Determination of Cell Concentration and Viability

1. D-PBS.
2. 0.05 % Trypsin–EDTA.
3. 0.1 % (v/v) Trypan blue solution in PBS.
4. Fuchs-Rosenthal hemocytometer (Brand, Wertheim, Germany) (*see* **Note 3**).

2.2.2 Virus Titration

1. Sterile flat bottom cell culture 24 wells microplates with lid (BD Falcon™ or equivalent).
2. Sterile tubes of 3.5 mL for viral dilutions and Flow cytometry analysis.
3. D-PBS.
4. Cell lysis solution: 10 mM Tris pH 8, 2 mM $MgCl_2$, 0.01 % (w/v) Triton X-100 (*see* **Note 4**).
5. DMEM supplemented with 10 % (v/v) fetal bovine serum, FBS (Gibco).
6. Flow cytometer instrument (PARTEC or equivalent).
7. HEK 293 cells (ATCC: CRL-1573) in DMEM with 10 % (v/v) FBS.
8. MDCK-E1 cells maintained in DMEM with 10 % (v/v) FBS
9. 0.25 % Trypsin–EDTA and 0.05 % Trypsin–EDTA.
10. HAdV and CAV-2 sample(s) to titrate (*see* **Note 4**).

2.3 Production in STB

2.3.1 Bioreactor Preparation

1. STBs with 2 and 5 L vessels (Biostat DCU-3 Bioreactors, Sartorius Stedium, Biotech GmbH or equivalent) (*see* **Note 5**).
2. Addition flasks for culture medium, cells, virus and harvesting.
3. Base solution: 1 M $NaHCO_3$.

2.3.2 HAdV Production

1. HEK 293 cells (ATCC: CRL-1573) adapted to suspension growth (*see* **Note 1**).
2. Ex-Cell® 293 serum-free medium (Sigma-Aldrich Co. LLC) supplemented with 4 mM of glutamine (Gibco) (*see* **Note 2**).
3. HEK 293 cells grown in bioreactor; cell concentration at infection (CCI) should be at 1×10^6 cells/mL.
4. Ex-Cell® 293 serum-free medium supplemented with 4 mM of glutamine.
5. Purified first generation HAdV vectors with GFP transgene or other gene of interest (*see* **Note 6**).

2.3.3 CAV-2 Production (See **Note 7**)

1. Washing solution: KOH in ethanol.
2. Dimethildichorosilane.
3. Toluene.
4. Cytodex™-1 microcarriers (GE Healthcare).
5. Inoculation flask.
6. D-PBS.
7. MDCK-E1 cells with 90 % confluence in T-Flasks.
8. Inoculation culture medium: Optipro™ SFM supplemented with 4 mM glutamine (*see* **Note 2**) and 5 % (v/v) FBS (*see* **Note 8**).

9. MDCK-E1 cells growing in microcarriers, after 2 days of inoculation, with final concentration of 1×10^6 cells/mL.
10. Optipro™ SFM supplemented with 4 mM glutamine (*see* **Note 2**).
11. Purified first generation CAV-2 with GFP transgene or other gene of interest (*see* **Note 6**).

3 Methods

3.1 Cell Culture

Cell manipulations are performed under sterile conditions in a laminar flow wood and cell-culture reagents are pre-warmed to 37 °C prior to use.

3.1.1 HEK 293 Cells Culture

HEK 293 cells suspension adapted are maintained in 500 mL Erlenmeyer flasks (80–100 mL working volume) and are split twice a week when attaining $2–3 \times 10^6$ cells/mL. The cells are kept inside an incubator at 37 °C with 8 % CO_2 in air in an orbital shaker at 130 rpm (*see* **Note 9**).

1. Take 0.5–1 mL of cell suspension sample from an Erlenmeyer with cells in culture for 3–4 days to an eppendorf tube.
2. Determine cell concentration (*see* Subheading 3.2.1).
3. Calculate the volume of cells necessary to inoculate an Erlenmeyer flask at $0.4–0.5 \times 10^6$ cells/mL with 80–100 mL.
4. Add the volume of cells suspension to a prefilled Erlenmeyer with the necessary volume of fresh Ex-Cell® 293 serum-free medium to attain the final working volume and resuspend well.
5. Place Erlenmeyer in incubator until next cell passage (3–4 days).

3.1.2 MDCK-E1 Cells Culture

MDCK-E1 cells are maintained in 150 cm^2 T-flask in an incubator at 37 °C, 5 % CO_2 in air. After reaching ~90 % confluence (twice a week), cells are passed to a new T-flask with a splitting ratio of 1:20.

1. Remove culture medium from the T-flask.
2. Rinse monolayer with 5 mL of D-PBS.
3. Remove D-PBS and add 5 mL of Trypsin–EDTA 0.25 %.
4. Place the T-Flask in the incubator until cell detachment becomes evident (~10 min).
5. Tap the side of the T-Flask to detach all cells.
6. Add 5 mL of Optipro™ SFM medium and resuspend cells.
7. Transfer cells suspension to a centrifugation tube and centrifuge the corresponding inoculum volume at $300 \times g$ for 10 min at room temperature.
8. Fill a new T-flask with 25 mL of fresh Optipro™ SFM medium.
9. Discard the supernatant of the centrifugation.

10. Resuspend pelleted cells with 5 mL fresh Optipro™ SFM medium.
11. Add cells suspension to the T-flask.
12. Place the T-flask in the incubator until new split (3–4 days).

3.2 Monitoring and Characterization of Cells and AdV Production

3.2.1 Determination of Cell Concentration and Viability

Cell concentration is determined by the trypan blue exclusion method counting cells in a hemocytometer (*see* **Note 3**). Cell suspensions are obtained directly from suspension cultures or after cell detachment (*see* Subheading 3.1) from adherent cultures.

Detachment of MDCK-E1 cells from microcarriers cultures is performed as follows:

1. Transfer 1 mL of cell culture to an eppendorf tube and allow microcarriers to settle-down.
2. Remove 800 μL of medium and wash cells with 800 μL PBS.
3. Remove PBS, add 800 μL Trypsin and incubate at 37 °C during 15 min. Agitate gently the eppendorf tube during this time (*see* **Note 10**)
4. Resuspend cells, let microcarriers settle-down and estimate cell concentration and viability with trypan blue dye using a hemocytometer.

3.2.2 Virus Titration

Quantification of infectious particles of both adenoviral vectors, done by monitoring the expression of GFP target cells (*see* **Note 11**), are shown as example. HEK 293 cells and MDCK-E1 cells used for the titration of infectious HAdV and CAV-2 vectors, respectively, are maintained in static conditions in DMEM supplemented with 10 % (v/v) FBS using a similar procedure described in Subheading 3.1.2. The FBS present in culture medium is sufficient to inactivate Trypsin, thus centrifugation step is not required when maintaining these cells.

1. Start with a 175 cm^2 T-flask with 80–90 % confluent HEK 293 cells (*see* **Note 12**) or MDCK-E1 cells (*see* **Note 13**).
2. Discard media and wash cells with 5 mL of D-PBS.
3. Discard D-PBS and add 4 mL of Trypsin–EDTA 0.05 %.
4. As soon as the cells start to detach, hit once on the side of the T-flask to detach the cells quicker.
5. Add 6 mL of DMEM with 10 % FBS and resuspend cells.
6. Transfer 0.5 mL to an eppendorf tube for cell counting.
7. Determine cell concentration (*see* Subheading 3.2.1).
8. Prepare a suspension of 0.25×10^6 HEK293 cells/mL, or 8×10^4 cells/mL MDCK-E1 cells in DMEM with 10 % FBS.
9. Seed cells in 24-well plates with 1 mL of this suspension per well (0.25×10^6 HEK293 cells/well or 4×10^4 MDCK-E1 cells/well).

10. Incubate at 37 °C with 5 % CO_2 for 24 h.
11. Discard media from two wells and wash cells carefully with 0.5 mL of D-PBS.
12. *For HEK293 cells*: Discard D-PBS and add 0.3 mL of Trypsin–EDTA 0.05 %. Incubate cells at room temperature until detachment is evident. Add 0.7 mL DMEM with 10 % FBS and resuspend the cells. *For MDCK-E1 cells*: Discard D-PBS and add 0.5 mL of Trypsin–EDTA 0.25 %. When detachment becomes evident add 0.5 mL DMEM 10 % FBS and resuspend the cells.
13. Transfer cells suspension to an eppendorf tube and determine cell concentration (*see* Subheading 3.2.1).
14. Dilute the viral suspension to be titrated on 3.5 mL tubes by serial dilutions from 10^{-1} to 10^{-6} (0.3 mL of viral suspension in 2.7 mL DMEM with 10 % FBS).
15. Discard media from each well of the cells.
16. Dispense 1 mL of the different viral dilutions onto the cells in a way to have two replicates for each dilution.
17. For at least two wells add just culture medium (negative control).
18. *For HEK293 cells*: Incubate at 37 °C with 5 % CO_2 for 17–20 h [21]. Transfer the content of each well to one 3.5 mL tube (*see* **Note 14**). Add 0.3 mL of Trypsin–EDTA 0.05 % to each well and incubate 30 s at room temperature. Add 0.7 mL DMEM with 10 % FBS and resuspend the cells. *For MDCK-E1 cells*: Incubate at 37 °C with 5 % CO_2 for 24 h. Collect cells from all wells by detaching them as stated in **steps 11–13**.
19. Transfer cells suspension to the respective 3.5 mL tube to be analyzed by flow cytometry and determine ip/mL according to Eq. 1 (*see* Subheading 3.2.4).

3.2.3 Flow Cytometry Analysis

Flow cytometric data is acquired using a CyFlow® Space and data analysis is performed using FlowMax® software from Partec. The green fluorescence signal is collected by a photomultiplier tube after passing through a 525 (±20) nm band pass filter (FL1).

1. Turn on the flow cytometer and make it ready for cell analysis.
2. Set the cytometric parameters to provide accurate discrimination between nonfluorescent negative cells and positive GFP-fluorescence cells on a FL1 versus FSC density plot using noninfected cells and highly infected cells as controls.
3. Start analyzing samples from the less to the higher infected.
4. Analyze GFP fluorescence of $>1 \times 10^4$ single viable cells per sample selected on FS versus SS scatter basis with high-fluorescent cells gated.
5. A minimum of two dilutions showing 3–30 % GFP-positive cells should be taken into account to calculate the viral titer.

3.2.4 Titer Calculation

To determine virus infectious titer the following equation (Eq. 1) must be used:

$$[\text{Titer}](\text{i.p.}/\text{mL}) = \frac{\%\ \text{GFP positive cells}}{\text{Volume of viral suspension}} \times \text{Viral Dilution} \times \text{Cell concentration at infection} \quad (1)$$

3.3 Production in STB

The protocols described below have been designed to produce 5 L of HAdV and 2 L CAV-2 using Biostat DCU-3 Bioreactors (*see* **Note 5**). The main points of bioreactor preparation are common for both cell culture systems used and are described in Subheading 3.3.1. The specific details for HAdV and CAV-2 are described in Subheadings 3.3.2 and 3.3.3, respectively. In Fig. 2 it is presented a process flow chart to help in the implementation of

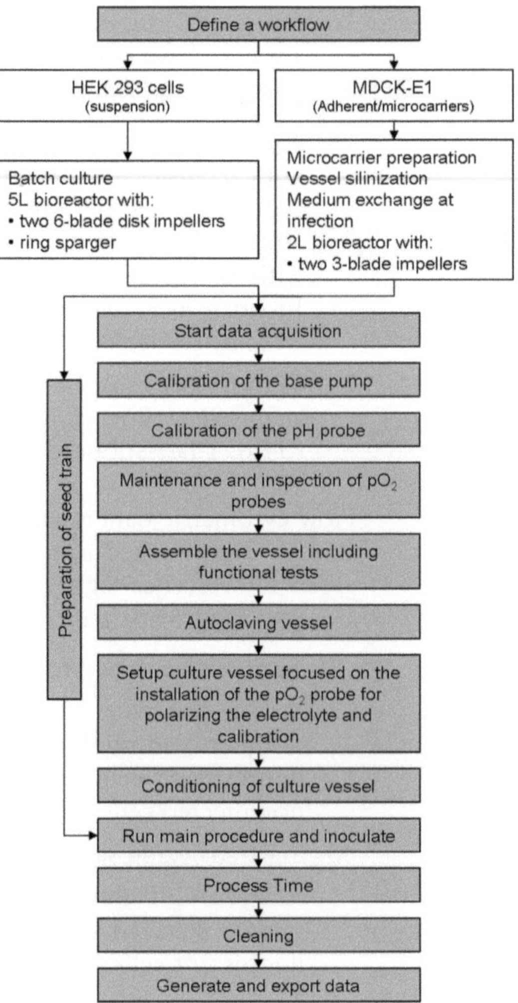

Fig. 2 Process flowchart

the protocols described in this heading. Table 1 represents the main operational and hydrodynamic conditions for cell culture and AdV infection in STB. All cell manipulations during bioreaction are performed under sterile conditions in a portable laminar flow wood. Parameters such as temperature, stirring speed, pH and pO_2 must be monitored and recorded throughout the all bioprocess including pH and pO_2 calibration (*see* **Note 15**).

3.3.1 Bioreactor Preparation

1. Calibrate pH probe with standard buffer solutions to pH 4.01 and pH 7.00 (*see* **Note 16**).
2. Check pO_2 probe response (*see* **Note 17**).
3. Set the aeration system with 0.2 μm vent filters for inlet gas flow (*see* **Note 18**): use direct sparging of air or oxygen with a ring sparger placed below the stirrer for the production of HAdV using HEK 293 cells, while for the production of CAV-2 using MDCK-E1 cells in microcarriers perform aeration through the headspace.
4. Before use, check that glass vessel is not damaged. To ensure the reactor is airtight, check all rubber rings and, if necessary, grease them.
5. Add MilliQ water to bioreactor vessel up to the level of probes after bioreactor assembly (*see* **Note 19**).
6. Assemble the bioreactor and tighten all fittings (*see* Fig. 3). Assemble the necessary impeller(s): 6-blade rushton turbine impellers for HEK 293 cells or 3-blade impellers for MDCK-E1 cells (*see* **Note 20**). Insert pH, pO_2, and temperature probes (*see* **Note 21**). Set exhaust cooler in place with 2×0.2 μm vent filters for outlet gas flow (*see* **Note 22**). Assemble all the bioreactor addition system and add entry connections for (*see* **Note 23**):
   (a) Culture medium.
   (b) Cell inoculation.
   (c) Virus infection/medium exchange at infection.
   (d) Extra entry (for safety).
7. Assemble the sampling/harvesting system with a sampling probe inside the bioreactor with silicone tube to reach the vessel base (*see* **Note 24**). Add an extra sampling probe to the bioreactor of CAV-2 vectors with proper height to perform the medium exchange at infection without removing the settled microcarriers. Connect the base bottle, secure all connections with cable ties and perform an hold-up test (*see* **Note 25**).
8. Prepare connecting bottles for medium, cell and virus inoculation in duplicates.
9. Autoclave bioreactor and connecting bottles at least 1 h, 121 °C for sterilization (*see* **Note 26**).

Table 1
Operational and hydrodynamic conditions for cell culture and AdV infection in stirred tank

Cell line	HEK 293	MDCK-E1
Culture type	Suspension cells	Adherent/microcarriers
Vessel characteristics		
System	BioStat DCU-3	
Vessel type	Univessel® 5 L Round bottom	Univessel® 2 L Round bottom
Total volume (L)	6.6	2.7
Working volume (L)	2–5	0.8–2
Impellers	Two 6-blade rushton turbine	Two 3-blade
Gassing system	Ring sparger	Headspace
Operational conditions		
Working volume (L)	5	2
pO_2 (%)	50[a]	40[b]
pH[c]	7.2 ± 0.05	7.4 ± 0.05
Agitation rate (rpm)	50–210	50
Aeration rate (vvm)	0.01	0.1
Temperature (°C)	37	
Inoculum (10^6 cells/mL)	0.5	0.2
CCI (10^6 cells/mL)	1	1
MOI (ip/cell)	5	5
TOH (hpi)	36–72	
Hydrodynamic conditions[d]		
Working volume (L)	5	2
Agitation rate (rpm)	50–210	50
Aeration rate (vvm)	0.01	0.1
Reynolds number (10^4)	0.48–20	0.27
Eddy size (μm)	25–73	181
Shear stress rate (N/m^2)	0.09–0.80	0.06
Tip speed (m/s)	0.17–0.70	0.14

[a] pO_2 is controlled by varying sequentially the nitrogen partial pressure in gas inlet, agitation rate, and oxygen partial pressure in gas inlet
[b] pO_2 is controlled by varying sequentially the nitrogen and oxygen partial pressure in gas inlet
[c] pH is controlled using CO_2 and 1 M $NaHCO_3$
[d] It is known that the hydrodynamics of the culture conditions are critical for scale-up since they affect cell growth and consequently virus production [33, 34]

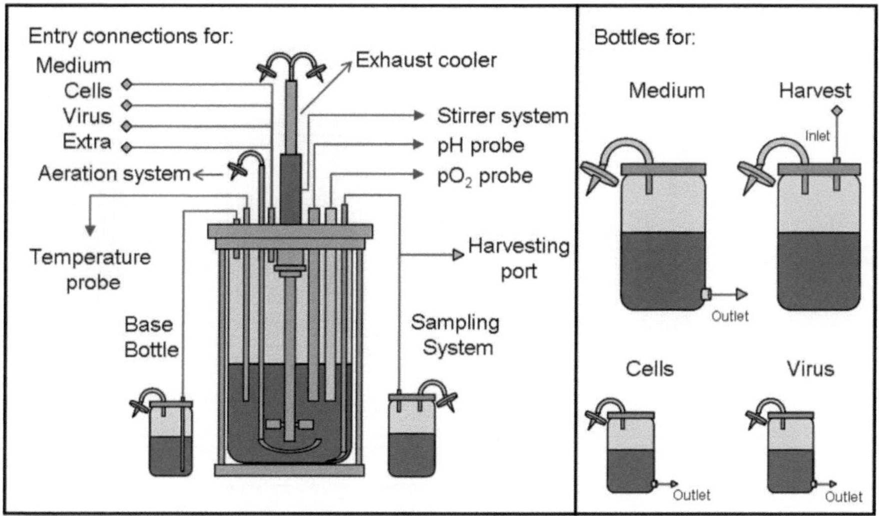

Fig. 3 Main components for bioreactor assembly. (Note: The base bottle and sampling system have steel caps with two integrated tube connectors)

10. Verify all bioreactor assembly and connections after sterilization (*see* **Note 27**).
11. Connect the bioreactor to the control unit as follows:
    (a) Water inlet and outlet to the jacket.
    (b) Cooler inlet and outlet to the control unit (*see* **Note 28**).
    (c) Temperature, pO_2, and pH probes.
    (d) Connect the motor.
    (e) Connect the aeration system to the gas supply system and start supplying N_2 to prepare pO_2 probe calibration.
    (f) When temperature is below 50 °C, fill the jacket with water and start temperature control (*see* **Note 29**).
12. Replace MilliQ water with 80 % working volume of culture medium (*see* **Note 30**).
    (a) Once temperature and pO_2 readings are stable, calibrate the pO_2 probe (*see* **Note 31**).
    (b) Fill the base bottle with the base solution and connect to the corresponding pump.
    (c) Select the desired pO_2 and pH and switch on the unit controllers (*see* **Note 32**).
    (d) When the desired pO_2 and pH are attained the bioreactor is ready for cell inoculation.

3.3.2 HAdV Production

The protocol to produce HAdV vectors is split into two points: (1) bioreactor inoculation and cell growth and (2) cell infection and virus harvesting. HEK 293 cells seed train can be prepared directly

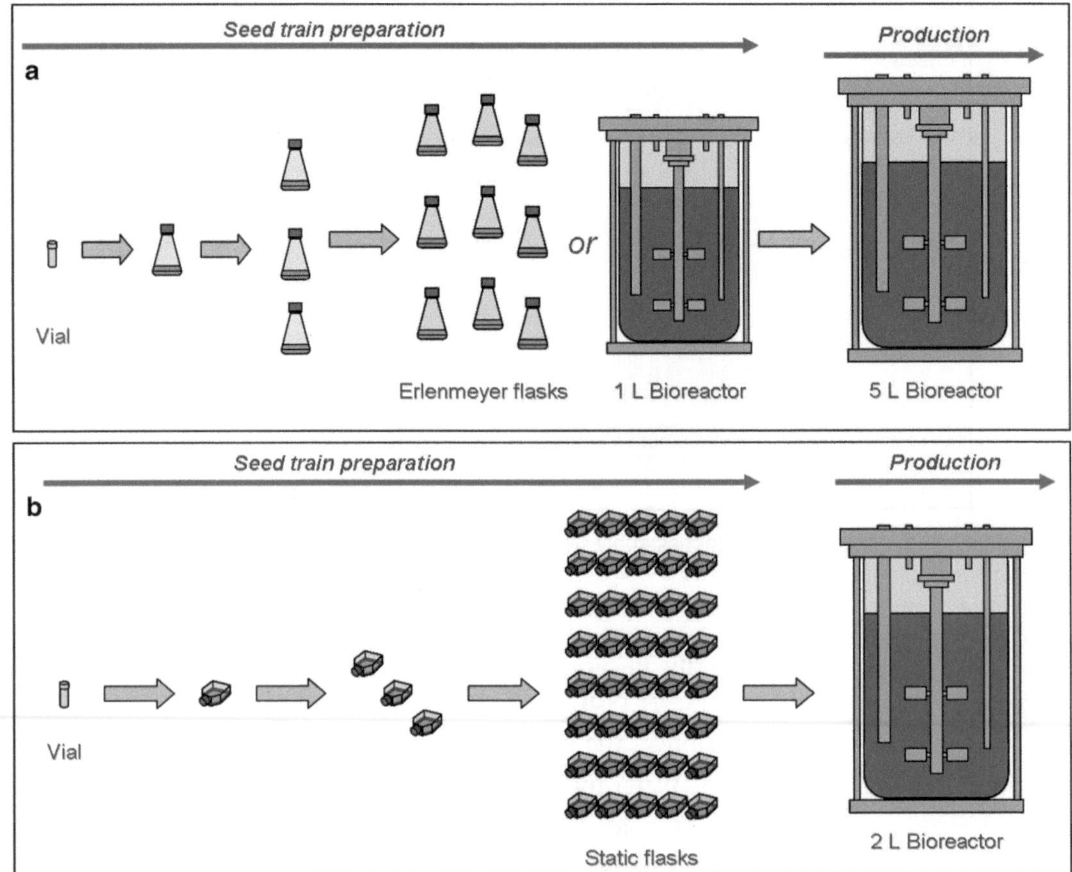

Fig. 4 Seed train strategies for: (**a**) HadV and (**b**) CAV-2 production

from cells grown in Erlenmeyer flasks (*see* Subheading 3.1.1) or from a previously inoculated 1 L bioreactor following Subheading 3.3.1 and adapted for 1 L working volume (*see* Fig. 4).

At this point, the bioreactor should have already 80 % of the culture medium (4 L) with the selected culture conditions stabilized (*see* Subheading 3.3.1).

1. Harvest cells from pre-inoculation vessels and determine cell concentration (*see* Subheading 3.2.1).
2. Centrifuge the corresponding volume of cells to ensure an inoculum of 5×10^5 cells/mL at $300 \times g$ for 10 min at room temperature (*see* **Note 33**).
3. Suspend cells in culture medium in order to bring the bioreactor volume to final working volume (20 % of working volume).
4. Confirm cell concentration and transfer cells to inoculation flask.
5. Connect the inoculation flask to the bioreactor and let them enter by gravity (*see* **Note 34**).

6. Monitor cell growth by sampling immediately after inoculation and at least every 24 h until the desired cell concentration to infect is reached. Cell infection can be performed when HEK 293 cells attain 1×10^6 cells/mL in batch mode.
7. Prepare infection flask with the corresponding amount of purified vectors (*see* **Note 6**) and fresh medium in order to infect cells with an MOI of five infectious particles per cell.
8. Connect the infection flask to the bioreactor and infect cells.
9. Sample the culture at least every 24 h for cell and virus analysis. Virus samples can be stored at −80 °C until analysis.
10. Harvest the culture bulk at 36–72 hpi (through the sampling system) creating pressure inside the vessel (*see* **Note 35**).

3.3.3 CAV-2 Production

Considering that MDCK-E1 cells are strictly adherent, microcarriers were used to develop a scalable stirred culture system. Thus, the production of CAV-2 vectors requires (1) silanization of glass material and (2) microcarriers preparation before (3) bioreactor inoculation and (4) cell infection and harvesting.

Glassware (bioreactor vessel, addition flasks, etc.) must be silanized to prevent microcarrier/cell adherence to the glass surface.

1. Fill the glass container with the washing solution and wait for at least 3 h and remove the solution (*see* **Note 7**).
2. Wash the glass container with water and let it dry completely (to evaporate the remaining water). If necessary, leave the glass container at 37 °C or overnight.
3. Add a small volume of dimethildichorosilane to the glass container(s) and rotate the container(s) to ensure the contact with the entire surface. Remove the excess solution (*see* **Note 7**).
4. Wash the glass container with a small volume of toluene and let it dry in the air protecting the container to avoid dust accumulation.
5. To ensure that the glass materials have been properly silanized, place a drop of water in the container: If the drop sticks to the glass, all the procedure must be repeated otherwise the material can be used.
6. Autoclave the container before use.

   Next steps are designed to prepare microcarriers for proper cell inoculation. In order to facilitate inoculation procedures, microcarriers are directly prepared in inoculation flask, used to insert cell inoculum inside bioreactor.

7. Weight 6 g of microcarrier beads (3 g/L) and transfer them to a silanized inoculation flask.
8. Suspend the microcarrier beads at 20 g/L in PBS and autoclave at 121 °C for 15 min, liquid cycle.

9. Allow the beads to settle-down, aspirate and discard the D-PBS and add the same volume of sterile D-PBS (final concentration 20 g/L). At this stage, and prior to use, microcarriers can be stored at 4 °C.
10. Prior to cell inoculation, aspirate and discard D-PBS, then rinse the beads twice in small amount (approximately 20 % of the final volume of inoculation flask) of culture medium.

    At this point, forty T-Flasks of 175 cm^2 (or equivalent) with 80–90 % of MDCK-E1 cells confluence are required (*see* Fig. 4), as well as the previously prepared microcarriers and bioreactor unit.
11. Harvest cells from standard cultures (*see* Subheading 3.1.2) and determine cell concentration using the trypan blue exclusion method (*see* Subheading 3.2.1).
12. Centrifuge the corresponding volume of cells to ensure an inoculum of 2×10^5 cells/mL at $300 \times g$ and during 10 min.
13. Suspend cells in inoculation culture medium (*see* Subheading 2.3.3) in order to bring the bioreactor volume to 100 % (2 L) of the working volume. Consider the volume occupied by microcarriers when adding the corresponding inoculation culture medium volume (*see* **Note 36**).
14. Confirm cell concentration and transfer cells to inoculation flask.
15. Connect the inoculation flask to the bioreactor and introduce the cells into the bioreactor by gravity, opening the clamps of the connected tubes (*see* **Note 34**).
16. Monitor cell growth by sampling and counting cells (*see* Subheading 3.2.1). When cells reach a concentration of 1×10^6 cells/mL, or 2 days after inoculation, infect cells to produce CAV-2 vectors.

    Once the concentration of 1×10^6 cells/mL is reached, cells can be infected. Culture medium is replaced during infection to maximize production of CAV-2 vectors.
17. Prepare medium/infection flask with fresh medium (Optipro™ SFM supplemented with 4 mM of glutamine) to exchange medium at infection. Consider the volume that can be removed without removing the microcarriers from the bioreactor to fill the medium/infection flask.
18. Add the corresponding amount of purified vectors (*see* **Note 6**) to the fresh medium in order to infect cells with an MOI of 5.
19. Perform the medium exchange/cell infection: stop bioreactor agitation, let microcarriers settle-down and remove medium using the extra sampling tube. Add the fresh medium with viral vectors and turn on the stirring.

20. Harvest the culture bulk at 36–72 hpi (through the sampling system) creating pressure inside the vessel (*see* **Note 35**). Alternatively separate intracellular and extracellular fractions by letting microcarriers settle-down and collecting supernatant to other container.

3.3.4 Bioreactor Disassembling

Prior to bioreactor disassembling and cleaning, sterilization is performed to inactivate any remaining viruses.

1. Switch off all the controllers.
2. Disconnect the cables and tubings connected to the control unit.
3. Remove the motor.
4. Prepare the bioreactor for sterilization to inactive viruses (*see* **Note 26**).
5. Autoclave the bioreactor.
6. Clean and store pO_2 and pH electrodes.
7. Remove the tubings installed on the top-plate and leave them overnight into lab disinfectants dissolved in warm water.
8. Wash the vessel and all components, rinse everything with 70 % ethanol and let it dry.
9. Store everything in the appropriated place until next use.

4 Notes

1. HEK 293 cells can be easily adapted to suspension to Ex-Cell® 293 serum-free medium or other commercially available cell culture media following media suppliers' instructions (directly or sequentially). MDCK-E1 cells were adapted sequentially to Optipro™ SFM medium following supplier instructions.
2. GlutaMAX™ media (Gibco) can be used in replacement to glutamine. It is a standard cell-culture media that contain a stabilized form of l-glutamine, the dipeptide l-alanyl-l-glutamine, that prevents degradation and ammonia build-up even during long-term cultures. Glutamine is normally used when the metabolism of the cells during growth and production is necessary to be evaluated since most metabolite analyzers have sensors for glutamine and not GlutaMAX™ media.
3. Several cell counting methods can be used including direct counting with hemocytometer or automatic cell counters. The method described herein uses Trypan blue, a dye that enters the cells whose membrane is damaged. In this way, when

analyzing a cell suspension at the microscope the viable cells appear bright and white, while dead cells appear blue, due to the presence of the dye in their cytoplasm. For samples analysis and cell counting under the microscope it is used a hemocytometer (Fuchs-Rosenthal or equivalent) that has two counting chambers with engraved squares to facilitate cell counting. The cell concentration is determined by Eq. 2:

$$\text{Cell concentration (cell / mL)} = \frac{\text{cells counted}}{\text{number of squares}} \times \text{dilution factor} \times \frac{1}{\text{square volume (cm}^3\text{)}} \quad (2)$$

where the dilution factor corresponds to the dilution of the sample in the dye and the square volume corresponds to square area multiplied by the chamber depth; this volume depends on the hemocytometer used.

4. The lysis solution is used to recover the intracellular viruses. Tris buffer is known to stabilize adenovirus vectors and the detergent Triton X-100 to cell lysis. Alternatively lyse cell with four freeze-cycles. Viral samples prior titrations are clarified at 3,000 × *g* for 10 min to remove debris that might interfere with infection process.
5. Glass bioreactors with working volumes ranging from 1 to 15 L are available from several sources. As these glass bioreactors are sterilized in autoclaves, the interior dimensions of the equipment have to be considered. For higher bioreactor volumes, most stirred tanks are made of stainless steal. In this case, sterilization is carried in situ using clean steam.
6. We recommend the use of a well-characterized viral seed stock if possible concentrated and purified either by standard ultracentrifugation procedures or chromatographic methods [22, 23]. One important issue to take into consideration is the storage conditions of the viral seed stock. Several factors, such as pH and temperature, affect the stability of the adenoviral vector preparations during storage [24, 25]. Several formulations have been developed for long-term storage in liquid and lyophilized form [26–28] from several months to more than 1 year at 2–8 °C (for more information *see* refs. 29–32).
7. The silanization ensures that the glass surface becomes less hydrophilic, allowing carriage, storage, and experiments with microcarriers. Not all microcarriers require silanization, so if you are not using Cytodex™, verify the manufacture recommendations. The washing solution for this procedure should be prepared as follows: dissolve 56 g of potassium hydroxide in 76 mL of distilled water with constant agitation in ice

(since it is an exothermic reaction); fill to 1,000 mL with 96 % ethanol and keep the solution at room temperature in ambar-glass bottle for 14 day before use. All the solution used in silanization protocol can be reused and should be stored in an ambar-glass container.

8. FBS is used during inoculation and initial cell growth to promote adherence of cells to microcarriers.
9. Stirring rates herein presented are only valid for orbital shaking motion with ⌀ of 10 mm (IKA model HS 260 or equivalent).
10. Use a thermomixer (37 °C, 300 rpm) to facilitate cell detachment.
11. The virus titration can be performed by several methods. Herein, a method for GFP expressing vectors is presented. For viral vectors without fluorescent reporters, infectious particles can be titrated using standard methods like $TCID_{50}$ or plaque assay. The total particles can be determined by HPLC, quantitative PCR, Nanosight or other, methods. More extensive characterization of the virus obtained is normally performed after purification.
12. One 175 cm^2 T-flask of 90 % confluence HEK 293 cells is sufficient for three to four 24-well plates. Each plate is sufficient to titrate two samples (six dilutions in duplicate).
13. One 175 cm^2 T-flask of 90 % confluence MDCK-E1 cells is sufficient for 8 plates of 24 wells.
14. HEK 293 cells detach very easily from the culture surface especially after infection. Careful is necessary when washing cells; alternatively, this step can be removed for infected cells. Collect also infected cells in suspension.
15. A supervisory control and data acquisition (SCADA) system should be used for monitor and control process values during preparation and run time, as exemplified in Table 1.
16. pH calibration is performed before the electrode is installed in the vessel using two commercial buffer of pH 4.01 and pH 7.00. Calibration determines the "zero drift" and "slope" of the electrode. First introduce the electrode inside the pH 7 buffer. Connect the pH and temperature sensors to the correct unit (where the bioreaction is going to run) to the electrode. Start the data acquisition for pH and temperature. When pH value is stable calibrate the "zero drift." Change to the pH 4 buffer and wait again for a stable pH value and calibrate the "slope." Make sure that the "zero drift" and "slope" values obtained are inside the limits of the electrodes used. If not, start the calibration again or check if electrode is still good for use.

17. The pO_2 electrode is calibrated in terms of percentage of oxygen saturation, after sterilization, with culture medium at operating conditions (temperature, airflow, and stirrer). To be sure that the electrode is good to be used is necessary to check its response. Connect the pH cable of the correct unit (where the bioreaction is going to run) to the electrode. Connect the temperature sensor cable to the unit. Observe the pO_2 value obtained with the electrode in the atmospheric air. Place the electrode in a container enriched with N_2 environment (can be obtained by supplying N_2 by the gas flow of the unit to the inside of a small cup covered with parafilm) and observe the decrease in pO_2 values of the electrode. A good electrode response corresponds to a decrease of pO_2 to 10 % in less than 1 min and to lower than 1 % in less than 10 min. If this criteria is not meet than verify the probe.
18. There are several types of aerations systems. In STBs with microcarrier cultures, cells at low densities or with low O_2 demanding, the aeration is performed via the headspace. The oxygen transfer in this case is affected by the surface area and the stirrer speed. The operation parameter ranges are 0.1–1 vvm. For higher cell densities and higher O_2 demanding cells, one of the systems used is the direct sparging of air or oxygen with ring spargers (drilled holes >0.5 mm diameter) placed below the stirrer. This allows better dispersion and longer residence times of gas bubbles in the culture liquid resulting in improved gas transfer rates. The operation ranges from 0.005 to 0.4 vvm.
19. The probes of the bioreactor should be submerged in liquid in order to avoid any damage by drying during sterilization. The liquid can be MilliQ water (resistivity of 18.2 MΩ cm) or D-PBS.
20. The number of the impellers assembled depends of the volume/size of the bioreactor and the type of the impeller on the type cell culture to be performed. The most common impellers for mammalian cells are 3-blade impellers (or marine impellers). These impellers allow less shear stress and are ideal for maintaining microcarriers in suspension. Other impellers also used are the rushton turbine with 4 or 6 bladed versions.
21. The probes should be placed separately and distributed along the bioreactor to avoid nonhomogeneous fluid distribution due to the electrodes interference. The pO_2 probe should be placed far from the entry ports since adding liquid to the bioreactor can create air bubbles that can be trapped under the electrode membrane an interfere with the measurements. If bubble trapping is observed, increase stirring rate for a moment to remove them.

22. The inlet filters (0.2 μm) placed in the exhaust cooler and aeration system serve to allow gas flow while maintaining the vessel under sterile conditions after sterilization. The exit of the exhaust cooler should have two filters attached. One is maintained open during sterilization to release pressure formed during sterilization. The other is maintained clamped as a backup if the other is damaged due to humidity release. The exhaust cooler should be placed on the same side of the water jacket tubes. In this way all the water connections are made in the same side of the bioreactor leaving the rest of the space free for easy handling all the other connections and sampling.
23. Use T associations or equivalent between connections, case the bioreactor ports are not enough to have all the connections necessary.
24. For the sampling system, it is necessary a sampling probe inside the bioreactor with silicone tubing at the bottom touching the vessel base. This will allow removing the water used during sterilization and can be also used as harvesting exit.
25. Perform a hold-up test to assure the entire bioreactor vessel and connecting bottles are properly sealed. Connect the gas supply to the air inlet filter of the aeration system and close all vessels exits. Start the airflow and set the pressure inside the vessel to 0.5 bar (this value is dependent on the vessel used). See if the airflow goes to zero when the desired pressure is reached inside the vessel and let the vessel under pressure a while to see if it is maintained. To depressurize the vessel, switch off the airflow and then open the tubing of exhaust cooler. Perform the hold-up test also in the sampling system and in the base bottle by connecting the gas supply to the inlet filter of the bottle. To be sure that the vessel is properly sealed, pressure should be maintained after the airflow reach zero for at least 1 h. If the airflow does not reach this value, check the sealing of all connections, probes, and vessel.
26. Before autoclaving the bioreactor, cover tubing, filter ends, probes, and stirrer system with aluminum foil and tape them with steam indicator tape. Clamp the aeration system if a sparger with direct contact to the water inside the bioreactor is connected, to avoid it exit during autoclaving. Open the clamp from the exhaust cooler to allow pressure release during sterilization.
27. After sterilization perform an hold-up test to make sure that all the connections are fit properly (some of the cable ties can be broken during autoclaving). If not, sterilize the vessel again.
28. The exhaust cooler has to be connected to the water supply as soon as possible to avoid the water condensation and exit filters damaging.

29. The temperature set point should be made in incremental steps in order to avoid the increase to very high temperatures.

30. The replacement of water by medium is made by removing the water by the sampling system. First the filter of the exhaust cooler is clamped to start pressurizing the vessel. Then the sampling system clamp is open so the water flows through the sampling bottle. The flask with medium is connected to the bioreactor and the medium is introduced by gravity. All the connections are made under sterile conditions.

31. Calibration requires the electrode polarization for about 6 h (connect the pO_2 cable to the electrode). Make sure that the "zero drift" and "slope" values obtained are inside the limits of the electrodes used. If not, start the calibration again.

32. The pH is controlled by aeration with a CO_2 gas mixture and 1 M $NaHCO_3$ added by a pre-calibrated feed pump. The base solution (1 M $NaHCO_3$) is prepared in MilliQ water and is sterilized by filtration (Autoclaving thermally degrade the solution). Other base solutions are sometimes also applied (KOH, NaOH, $NaHCO_2$, or Na_2CO_2) depending on the buffer system of the culture medium.

33. For larger scales, since the inoculum corresponds normally a dilution 1:6 (0.5×10^6–3×10^6 cells/mL) of the starting culture, the vessel inoculation can be performed directly without centrifugation.

34. The inoculation should be gentle to avoid cell damage and the flask should be agitated in order to avoid deposition of cells. If inoculum is not being transferred through gravity pump for 2–3 s air through the filter inlet of inoculation flask. After inoculation, the jacket temperature normally increases considerable to compensate the decrease in temperature inside the reactor. To overcome this problem, add cooling water manually by pressing the switch for water supply. Alternatively, turn off the temperature control during inoculation and turn back on afterwards controlling the temperature back to 37 °C in a slow rate.

35. Harvesting time depends on the downstream process purification and corresponding working volume. Low-volume working volumes should be harvested up to 48 hpi, and centrifuged to concentrate viruses by collecting only intracellular fraction. Alternatively, if downstream process is managed to high working volumes, culture bulk can be harvested latter, without this concentration step.

36. As a guide to calculate the settle volume and medium entrapment by microcarriers, 1 g of Cytodex™, for example, has a volume of 15–18 mL.

Acknowledgments

The authors acknowledge the financial support from Fundação para a Ciência e Tecnologia (FCT), Portugal (projects PTDC/EBB-BIO/119501/2010 and PTDC/EBB-BIO/118615/2010) and the FP7 EU project BrainCAV (HEALTH-HS_2008_222992). A.C. Silva and P. Fernandes acknowledge the FCT for the Ph.D. grants SFRH/BD/45786/2008 and SFRH/BD/70810/2010, respectively.

References

1. McConnell MJ, Imperiale MJ (2004) Biology of adenovirus and its use as a vector for gene therapy. Hum Gene Ther 15(11):1022–1033
2. Tatsis N, Ertl HC (2004) Adenoviruses as vaccine vectors. Mol Ther 10(4):616–629
3. Dormond E, Perrier M, Kamen A (2009) From the first to the third generation adenoviral vector: what parameters are governing the production yield? Biotechnol Adv 27(2):133–144
4. Sheets RL, Stein J, Bailer RT, Koup RA, Andrews C, Nason M, He B, Koo E, Trotter H, Duffy C, Manetz TS, Gomez P (2008) Biodistribution and toxicological safety of adenovirus type 5 and type 35 vectored vaccines against human immunodeficiency virus-1 (HIV-1), Ebola, or Marburg are similar despite differing adenovirus serotype vector, manufacturer's construct, or gene inserts. J Immunotoxicol 5(3):315–335
5. Shott JP, McGrath SM, Pau MG, Custers JH, Ophorst O, Demoitie MA, Dubois MC, Komisar J, Cobb M, Kester KE, Dubois P, Cohen J, Goudsmit J, Heppner DG, Stewart VA (2008) Adenovirus 5 and 35 vectors expressing Plasmodium falciparum circumsporozoite surface protein elicit potent antigen-specific cellular IFN-gamma and antibody responses in mice. Vaccine 26(23): 2818–2823
6. Kremer EJ, Boutin S, Chillon M, Danos O (2000) Canine adenovirus vectors: an alternative for adenovirus-mediated gene transfer. J Virol 74(1):505–512
7. Perreau M, Kremer EJ (2005) Frequency, proliferation, and activation of human memory T cells induced by a nonhuman adenovirus. J Virol 79(23):14595–14605
8. Soudais C, Boutin S, Kremer EJ (2001) Characterization of cis-acting sequences involved in canine adenovirus packaging. Mol Ther 3(4):631–640
9. Bru T, Salinas S, Kremer EJ (2010) An update on canine adenovirus type 2 and its vectors. Viruses 2(9):2134–2153
10. Fallaux FJ, Bout A, van der Velde I, van den Wollenberg DJ, Hehir KM, Keegan J, Auger C, Cramer SJ, van Ormondt H, van der Eb AJ, Valerio D, Hoeben RC (1998) New helper cells and matched early region 1-deleted adenovirus vectors prevent generation of replication-competent adenoviruses. Hum Gene Ther 9(13):1909–1917
11. Schiedner G, Hertel S, Kochanek S (2000) Efficient transformation of primary human amniocytes by E1 functions of Ad5: generation of new cell lines for adenoviral vector production. Hum Gene Ther 11(15):2105–2116
12. Cote J, Garnier A, Massie B, Kamen A (1998) Serum-free production of recombinant proteins and adenoviral vectors by 293SF-3F6 cells. Biotechnol Bioeng 59(5):567–575
13. Nadeau I, Gilbert PA, Jacob D, Perrier M, Kamen A (2002) Low-protein medium affects the 293SF central metabolism during growth and infection with adenovirus. Biotechnol Bioeng 77(1):91–104
14. Ferreira TB, Ferreira AL, Carrondo MJ, Alves PM (2005) Two different serum-free media and osmolality effect upon human 293 cell growth and adenovirus production. Biotechnol Lett 27(22):1809–1813
15. Maranga L, Aunins JG, Zhou W (2005) Characterization of changes in PER.C6 cellular metabolism during growth and propagation of a replication-deficient adenovirus vector. Biotechnol Bioeng 90(5):645–655
16. Mendonca RZ, Prado JCM, Pereira CA (1999) Attachment, spreading and growth of VERO cells on microcarriers for the optimization of large scale cultures. Bioprocess Biosyst Eng 20:565–571
17. Wu SC, Huang GY, Liu JH (2002) Production of retrovirus and adenovirus vectors for gene

therapy: a comparative study using microcarrier and stationary cell culture. Biotechnol Prog 18(3):617–622

18. Kamen A, Henry O (2004) Development and optimization of an adenovirus production process. J Gene Med 6(Suppl 1):S184–S192
19. Silva AC, Peixoto C, Lucas T, Kuppers C, Cruz PE, Alves PM, Kochanek S (2010) Adenovirus vector production and purification. Curr Gene Ther 10(6):437–455
20. Fernandes P, Peixoto C, Santiago VM, Kremer EJ, Coroadinha AS, Alves PM (2013) Bioprocess development for canine adenovirus type 2 vectors. Gene Ther 20:353–360
21. Sandhu KS, Al-Rubeai M (2008) Monitoring of the adenovirus production process by flow cytometry. Biotechnol Prog 24(1):250–261
22. Segura MM, Puig M, Monfar M, Chillon M (2012) Chromatography purification of canine adenoviral vectors. Hum Gene Ther Methods 23:182–197
23. Peixoto C, Ferreira TB, Sousa MF, Carrondo MJ, Alves PM (2008) Towards purification of adenoviral vectors based on membrane technology. Biotechnol Prog 24(6):1290–1296
24. Croyle MA, Roessler BJ, Davidson BL, Hilfinger JM, Amidon GL (1998) Factors that influence stability of recombinant adenoviral preparations for human gene therapy. Pharm Dev Technol 3(3):373–383
25. Obenauer-Kutner LJ, Ihnat PM, Yang TY, Dovey-Hartman BJ, Balu A, Cullen C, Bordens RW, Grace MJ (2002) The use of field emission scanning electron microscopy to assess recombinant adenovirus stability. Hum Gene Ther 13(14):1687–1696
26. Croyle MA, Cheng X, Wilson JM (2001) Development of formulations that enhance physical stability of viral vectors for gene therapy. Gene Ther 8(17):1281–1290
27. Evans RK, Nawrocki DK, Isopi LA, Williams DM, Casimiro DR, Chin S, Chen M, Zhu DM, Shiver JW, Volkin DB (2004) Development of stable liquid formulations for adenovirus-based vaccines. J Pharm Sci 93(10):2458–2475
28. Cruz PE, Silva AC, Roldao A, Carmo M, Carrondo MJ, Alves PM (2006) Screening of novel excipients for improving the stability of retroviral and adenoviral vectors. Biotechnol Prog 22(2):568–576
29. Rexroad J, Wiethoff CM, Green AP, Kierstead TD, Scott MO, Middaugh CR (2003) Structural stability of adenovirus type 5. J Pharm Sci 92(3):665–678
30. Altaras NE, Aunins JG, Evans RK, Kamen A, Konz JO, Wolf JJ (2005) Production and formulation of adenovirus vectors. Adv Biochem Eng Biotechnol 99:193–260
31. Rexroad J, Evans RK, Middaugh CR (2006) Effect of pH and ionic strength on the physical stability of adenovirus type 5. J Pharm Sci 95(2):237–247
32. Rexroad J, Martin TT, McNeilly D, Godwin S, Middaugh CR (2006) Thermal stability of adenovirus type 2 as a function of pH. J Pharm Sci 95(7):1469–1479
33. Cruz PE, Cunha A, Peixoto CC, Clemente J, Moreira JL, Carrondo MJ (1998) Optimization of the production of virus-like particles in insect cells. Biotechnol Bioeng 60(4):408–418
34. Maranga L, Cunha A, Clemente J, Cruz P, Carrondo MJ (2004) Scale-up of virus-like particles production: effects of sparging, agitation and bioreactor scale on cell growth, infection kinetics and productivity. J Biotechnol 107(1):55–64

Chapter 14

Canine Adenovirus Downstream Processing Protocol

Meritxell Puig, Jose Piedra, Susana Miravet, and María Mercedes Segura

Abstract

Adenovirus vectors are efficient gene delivery tools. A major caveat with vectors derived from common human adenovirus serotypes is that most adults are likely to have been exposed to the wild-type virus and exhibit active immunity against the vectors. This preexisting immunity limits their clinical success. Strategies to circumvent this problem include the use of nonhuman adenovirus vectors. Vectors derived from canine adenovirus type 2 (CAV-2) are among the best-studied representatives. CAV-2 vectors are particularly attractive for the treatment of neurodegenerative disorders. In addition, CAV-2 vectors have shown great promise as oncolytic agents in virotherapy approaches and as vectors for recombinant vaccines. The rising interest in CAV-2 vectors calls for the development of scalable GMP compliant production and purification strategies. A detailed protocol describing a complete scalable downstream processing strategy for CAV-2 vectors is reported here. Clarification of CAV-2 particles is achieved by microfiltration. CAV-2 particles are subsequently concentrated and partially purified by ultrafiltration–diafiltration. A Benzonase® digestion step is carried out between ultrafiltration and diafiltration operations to eliminate contaminating nucleic acids. Chromatography purification is accomplished in two consecutive steps. CAV-2 particles are first captured and concentrated on a propyl hydrophobic interaction chromatography column followed by a polishing step using DEAE anion exchange monoliths. Using this protocol, high-quality CAV-2 vector preparations containing low levels of contamination with empty viral capsids and other inactive vector forms are typically obtained. The complete process yield was estimated to be 38–45 %.

Key words CAV-2 vectors, Purification, Membrane filtration, Chromatography

1 Introduction

A number of viruses have been used for gene delivery purposes. Among them, human adenovirus vectors are the most commonly used vectors in clinical trials (http://www.wiley.com//legacy/wileychi/genmed/clinical/). Adenovirus vectors possess several key advantages including their high production titers, high transduction efficiency in both replicating and differentiated cells, large cloning capacity and non-integrative nature. However, their clinical utility is severely compromised by the fact that most patients are likely to have been exposed to wild-type human adenoviruses and possess active immunity against common adenovirus vectors

Miguel Chillón and Assumpció Bosch (eds.), *Adenovirus: Methods and Protocols*, Methods in Molecular Biology, vol. 1089,
DOI 10.1007/978-1-62703-679-5_14, © Springer Science+Business Media, LLC 2014

derived from human serotypes 2 and 5 [1, 2]. Strategies to circumvent this problem include the use of less prevalent human serotypes or adenovirus vectors derived from nonhuman adenoviruses.

Canine adenovirus vectors type 2 (CAV-2) are most likely the best-studied representative of nonhuman adenovirus vectors. These vectors retain key advantages of human Ad5; yet, despite intimate cohabitation of humans and dogs, CAV-2 has not been shown to cross the species barrier and cannot replicate in human cells [3]. Upon injection in the brain, CAV-2 vectors exhibit preferential tropism for neurons and have shown a high level of retrograde axonal transport that allows them to reach cells located at distant brain regions [4]. These characteristics make them particularly attractive for targeting neurodegenerative diseases using gene therapy approaches [5]. In addition, CAV-2 vectors are being evaluated as oncolytic agents in virotherapy approaches [6–8] and as oral vaccines for the control of rabies in wild and domestic animals [9–13].

The rising interest in CAV-2 vectors for gene therapy, oncolytic therapy and vaccine approaches calls for the development of scalable GMP compliant production and purification strategies as significant amounts of clinical-grade material are required for preclinical and clinical trials [14]. Recently, scalable downstream processing strategies specifically tailored for CAV-2 have been reported [15, 16]. These strategies combine multiple membrane filtration and chromatography purification steps.

A detailed step-by-step protocol for the purification of CAV-2 vectors is reported here. Primary recovery of CAV-2 particles is achieved by microfiltration and ultrafiltration–diafiltration. A nuclease digestion step is carried out between ultrafiltration and diafiltration operations. Chromatography purification is accomplished by capture of CAV-2 vectors using hydrophobic interaction chromatography followed by CAV-2 vector polishing by anion exchange chromatography. A representative scheme indicating the yield that can be expected along the process is presented (*see* Fig. 1). Of note, the higher the number of downstream processing steps, the higher the purity achieved but the lower the virus yield. Thus, the required level of purity needs to be determined in advance according to the application being considered. The complete strategy renders high-quality CAV-2 vector preparations with low contamination with empty viral capsids and other inactive vector forms as judged by total and infective particle quantitation assays, $OD_{260/280}$ chromatography absorbance ratios, electron microscopy and electrophoretic analyses [16]. The complete process overall yield was estimated to be between 38 and 45 % [16], which is well in line with those reported for other viral vectors [17, 18].

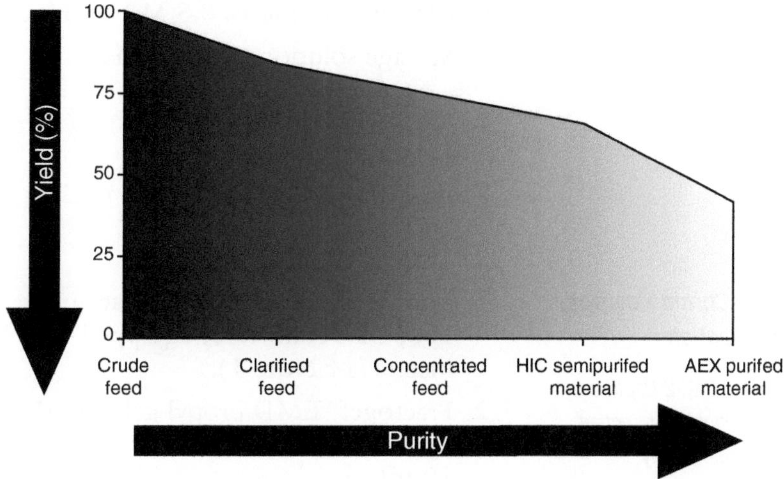

Fig. 1 CAV-2 downstream processing. Crude CAV-2 vector stocks are subjected to a series of DSP steps in order to attain the desired level of purity. With each additional step vector purity considerably improves but the overall process yield declines as shown in the scheme

2 Materials

2.1 Clarification by Dead-End Microfiltration

1. Peristaltic pump, silicone tubings, and pressure gauge.
2. ULTA™ cap PP membrane (0.6 μm) capsule filtration device (GE Healthcare, Uppsala, Sweden).
3. Milli-Q H_2O.
4. Buffer A: 20 mM Tris–HCl, 2 mM $MgCl_2$, 10 mM NaCl, 2.5 % glycerol, pH 8 filter-sterilized buffer.

2.2 Concentration by Cross-Flow Ultrafiltration

2.2.1 Midjet® System (Up to 200 mL)

1. Advanced Midjet® system (GE healthcare).
2. Hollow fiber MidGee® cartridge, MWCO 100,000 (GE Healthcare).
3. Connecting silicone tubing (size 14).
4. Diafiltration buffer: 20 mM Tris–HCl, 2 mM $MgCl_2$, 10 mM NaCl, 2.5 % glycerol, pH 8 filter-sterilized buffer.
5. Cleaning solutions: 0.5 M NaOH, 20 % EtOH, Milli-Q H_2O.
6. Storage solution: 20 % EtOH.

2.2.2 Quixstand® System (Up to 10 L)

1. Quixstand® system (GE healthcare) and Watson Marlow peristaltic pump.
2. Hollow fiber Xampler® cartridge, MWCO 100,000 (GE Healthcare).
3. Connecting tubing (size 18).
4. Diafiltration buffer: 20 mM Tris–HCl, 2 mM $MgCl_2$, 10 mM NaCl, 2.5 % glycerol, pH 8 filter-sterilized buffer.

5. Cleaning solutions: 0.5 M NaOH, 20 % EtOH, Milli-Q H_2O.
6. Storage solution: 20 % EtOH.

2.3 Benzonase® Treatment

1. Benzonase grade I 25,000 U (Merck Millipore, Darmstadt, Germany).
2. Digestion buffer: 100 mM Tris–HCl, 100 mM $MgCl_2$, pH 8 filter-sterilized buffer.

2.4 Chromatography Purification

2.4.1 CAV-2 Capture by Hydrophobic Interaction Chromatography

1. Low-pressure liquid chromatography system (AKTA explorer 100; GE Healthcare) equipped with UV, conductivity and pH meters (*see* **Note 1**).
2. Fractogel® EMD propyl gel (S) (Merck Millipore) packed into a XK 16/20 column (GE Healthcare) to a final volume of 10-mL.
3. 0.45 μm pore size Millex-HV PVDF syringe-mounted filters (Merck Millipore).
4. Buffer A: 20 mM Tris–HCl, 2 mM $MgCl_2$, 10 mM NaCl, 2.5 % glycerol, pH 8 (*see* **Note 2**).
5. Buffer B: 2 M $NH_4(SO_4)_2$ in 20 mM Tris–HCl, 2 mM $MgCl_2$, 10 mM NaCl, 2.5 % glycerol, pH 8.
6. Storage solution: 20 % EtOH + 150 mM NaCl in Milli-Q H_2O (*see* **Note 3**).
7. Cleaning solution: 0.5 M NaOH.

2.4.2 Polishing of CAV-2 Using Anion Exchange Chromatography Monoliths

1. Low-pressure liquid chromatography system (AKTA explorer 100; GE Healthcare) equipped with UV, conductivity and pH meters.
2. Convective Interaction Media (CIM®) DEAE monolithic disks (0.34 mL) (Bia Separations, Ljubljana, Slovenia) (*see* **Note 4**).
3. Buffer A: 20 mM Tris–HCl, 2 mM $MgCl_2$, 10 mM NaCl, 2.5 % glycerol, pH 7.
4. Buffer B: 2 M NaCl in 20 mM Tris–HCl, 2 mM $MgCl_2$, 10 mM NaCl, 2.5 % glycerol, pH 7.
5. Storage solution: 20 % EtOH.
6. Cleaning solution: 1 M NaOH.

3 Methods

3.1 Clarification by Dead-End Microfiltration

1. Thaw CAV-2 virus stocks using water bath at 37 °C (*see* **Notes 5** and **6**).
2. Aliquot starting material samples for analyses and measure the starting volume. Keep the viral stock at 4 °C until further processing.

3. Set up the filtration system in the biological safety cabinet (*see* **Note 7**).
4. Filter viral stocks at constant flow rate (start at 10 mL/min and gradually increase up to 50 mL/min). Collect the permeate in a sterile collection bottle. Check pressure at all times (*see* **Note 8**).
5. Measure the final volume and aliquot clarified virus stock samples for analyses (*see* **Note 9**).
6. Keep the virus stock at 4 °C until the next downstream processing step or store at −80 °C along with the aliquoted samples.

3.2 Concentration by Cross-Flow Ultrafiltration–Diafiltration

3.2.1 Midjet® System (Up to 200 mL)

1. Set up the ultrafiltration system in the biological safety cabinet (*see* **Note 10**).
2. Recirculate 200 mL of 20 % EtOH through new cartridges during 10 min to remove trace amounts of glycerol humectants. Then clean cartridges by recirculating with 200 mL Milli-Q H_2O during 15 min and flush the cartridge membrane with 200 mL of diafiltration buffer A. The membrane should be kept wet at all times from this point on (*see* **Note 11**).
3. Transfer the virus stock into a clean reservoir (max. volume 200 mL). Aliquot starting material samples for analyses and measure the starting volume.
4. Connect the feed reservoir and start pumping sample through the cartridge. Adjust the transmembrane pressure (TMP) to attain a recirculation flow of 60 liters per square meter of membrane area per hour (LMH) using the backpressure valve attached to the outlet tubing (*see* **Note 12**).
5. Once the volume is reduced by tenfold, empty tubing in order to collect the entire sample in the feed reservoir. This can be accomplished by disconnecting the inlet tubing from the reservoir cap.
6. Follow the procedure described in Subheading 3.3 to digest nucleic acids inside the feed reservoir in between the ultrafiltration and diafiltration operations.
7. Diafilter the sample by adding into the feed reservoir a volume of diafiltration buffer equal to that of the retentate. Repeat diafiltration three times in discontinuous mode.
8. The ultrafiltration–diafiltration process is stopped once the final desired volume of retentate is reached after the third diafiltration step. Empty the tubing in order to collect the entire sample in the feed reservoir.
9. Measure the final volume and aliquot concentrated virus stock samples for analyses (*see* **Note 13**)

10. Store concentrated samples at −80 °C along with the previously aliquoted samples.
11. Clean cartridge by recirculating NaOH 0.5 M during 20 min. Then neutralize with 200 mL of buffer A and 200 mL of Milli-Q H_2O. Store in 20 % EtOH at 4 °C.

3.2.2 Quixstand® System (Up to 10 L)

1. Set up the ultrafiltration system in the biological safety cabinet (*see* **Note 10**).
2. Recirculate 500 mL of 20 % EtOH through new cartridges during 10 min to remove trace amounts of glycerol humectants. Then clean cartridges by recirculating with 500 mL Milli-Q H_2O during 15 min and flush the cartridge membrane with 500 mL of diafiltration buffer A. The membrane should be kept wet at all times from this point on (*see* **Note 11**).
3. Transfer the virus stock into a clean autoclavable reservoir (max. volume 1,000 mL). Connect the system to a secondary feed reservoir if volumes larger than 1 L are being processed. Aliquot starting material samples for analyses and measure the starting volume.
4. Start pumping sample at 400 rpm through the cartridge. Adjust the transmembrane pressure (TMP) to attain a recirculation flow of 60 LMH using the backpressure valve attached to the outlet tubing (*see* **Note 12**).
5. Once the volume is reduced by tenfold, empty tubing in order to collect the entire sample in the feed reservoir. This can be accomplished by disconnecting the inlet tubing from the reservoir cap.
6. Follow the procedure described in Subheading 3.3 to digest nucleic acids inside the feed reservoir in between the ultrafiltration and diafiltration operations.
7. Diafilter the sample by adding into the feed reservoir a volume of diafiltration buffer equal to that of the retentate. Repeat diafiltration three times in discontinuous mode.
8. The ultrafiltration–diafiltration process is stopped once the final desired volume of retentate is reached after the 3rd diafiltration step. Empty the tubing in order to collect the entire sample in the feed reservoir.
9. Measure the final volume and aliquot concentrated virus stock samples for analyses (*see* **Note 13**).
10. Store concentrated samples at −80 °C along with the previously aliquoted samples.
11. Clean cartridge by recirculating NaOH 0.5 M during 20 min. Then neutralize with 200 mL of buffer A and 200 mL of Milli-Q H_2O. Store in 20 % EtOH at 4 °C.

3.3 Benzonase® Treatment

1. Prepare a working solution (WS) stock of Benzonase® in digestion buffer containing 10,000 U/mL just prior to digestion (*see* **Note 14**).
2. Once all the sample is in the feed reservoir (including the volume inside the filter and connections), measure the volume of retentate reached.
3. Treat the concentrated virus stock with 100 U/mL of Benzonase® by adding 10 μL of WS/mL of sample into the feed reservoir. Mix by recirculating sample 3 times for 15 s and let digestion go for 60 min at room temperature (*see* **Note 15**). A representative example of digestion performance using either the Midjet® or Quixstand® systems is shown in Fig. 2.
4. When the time is over continue with diafiltration process as described in Subheading 3.2.1 or Subheading 3.2.2.

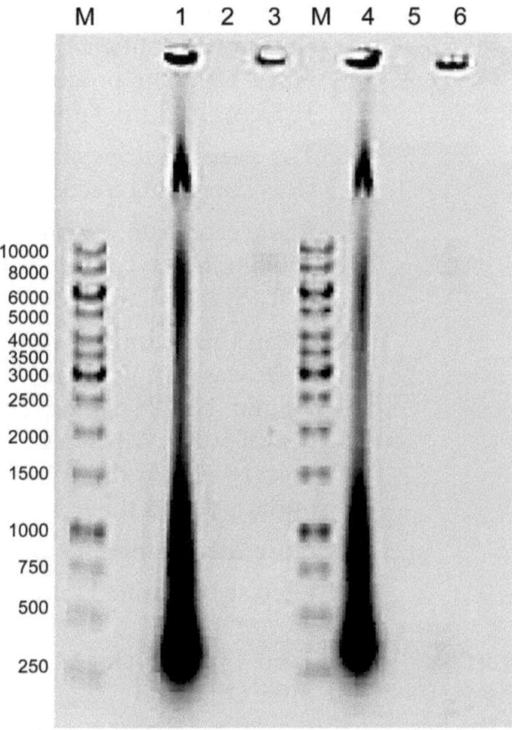

Fig. 2 Benzonase® digestion. Agarose 1 % gel. Lanes: (M) Molecular weight marker (GeneRuler™ 1,000 bp DNA ladder, Fermentas). Retentate (1), permeate (2), and Benzonase® treated retentate (100 U/mL, 1 h at room temperature) (3) using the Midjet® system. Retentate (4), permeate (5), and Benzonase® treated retentate (100 U/mL, 1 h at room temperature) (6) using the Quixstand® system

3.4 Chromatography Purification

3.4.1 CAV-2 Capture by Hydrophobic Interaction Chromatography

1. Turn on the ÄKTA explorer and wash pump A & B with Milli-Q H_2O
2. Install the 10-mL Fractogel® propyl column and remove the storage buffer with 10 column volumes (CV) of Milli-Q H_2O at a linear flow rate of 92 cm/h (3 mL/min).
3. Wash pump A and B with the corresponding buffers. Monitor UV absorbance at 280 and 260 nm (*see* **Note 16**). Equilibrate the column with binding buffer (0.85 M $NH_4(SO_4)_2$ in buffer A).
4. Thaw the tenfold concentrated CAV-2 stock. Dilute sample in buffer A containing 1.7 M $NH_4(SO_4)_2$ to match the conductivity of the binding buffer (120 mS/cm) (typically 1:2). Filter the conditioned chromatography feed using a 0.45-μm pore size syringe-mounted filters. Aliquot starting material for analyses (*see* **Note 17**).
5. When stable baseline is achieved, load the conditioned CAV-2 feed (*see* **Note 18**) and apply a step-wise gradient elution strategy that includes a wash step in binding buffer (5 CV) to remove the bulk of contaminating proteins, followed by CAV-2 elution with buffer A using a single-step gradient (16 CV) (*see* **Note 19**). The process is carried out at room temperature at 153 cm/h (5 mL/min).
6. The virus particles elute in a defined peak (*see* Fig. 3a). Pool virus-containing fractions and aliquot for analyses (*see* **Note 20**).
7. After each run, wash the column with buffer A and re-equilibrate the column with binding buffer containing 0.85 M $NH_4(SO_4)$.
8. Clean the column with 0.5 M NaOH every three runs. Load on at least 3 CV of the cleaning solution and let sit for at least 20 min. NaOH can be removed by rinsing the column with 10 CV of 0.15 M NaCl in Milli-Q H_2O. Re-equilibrate with at least 10 CV of the equilibration buffer. Check pH at column outlet prior to the next run.
9. Store the column in storage buffer with 10 CV at 92 cm/h.
10. Keep semi-purified samples at 4 °C until further use.

3.4.2 Polishing of CAV-2 Using Anion Exchange Chromatography Monoliths

1. Turn on the ÄKTA explorer and wash pump A & B with Milli-Q H_2O
2. Install the CIM® DEAE monolithic disk and remove the storage buffer with 10 CV of Milli-Q H_2O at a flow rate of 2 mL/min.
3. Wash pump A and B with the corresponding buffers. Monitor UV absorbance at 280 and 260 nm. Equilibrate the column with binding buffer to attain a stable baseline. Binding buffer consists of 160 mM NaCl in buffer A.

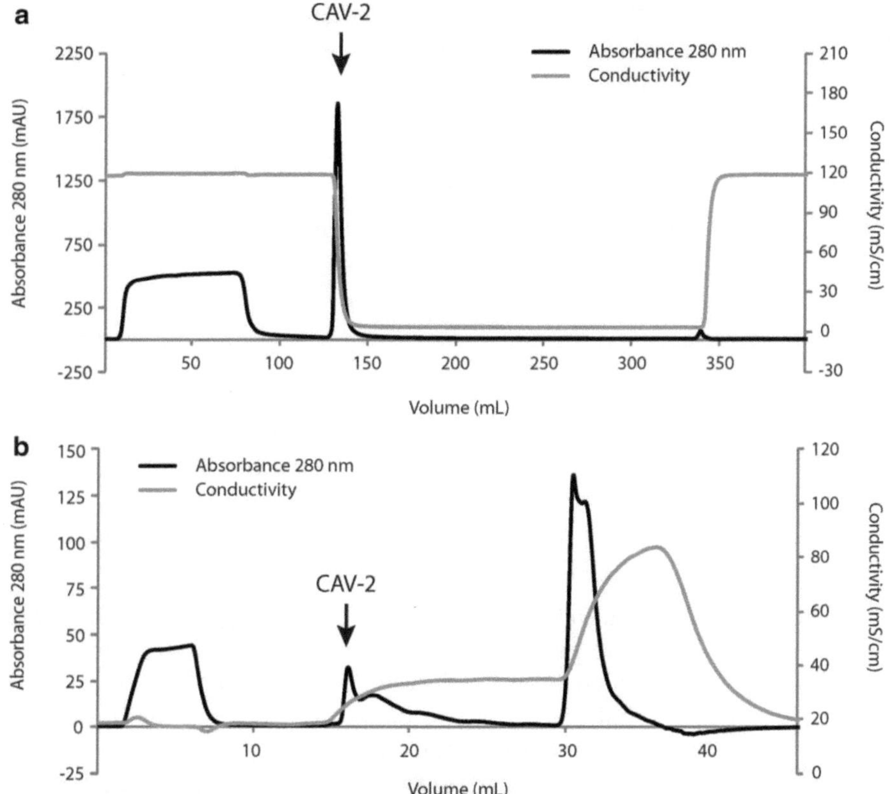

Fig. 3 HIC and AEX chromatography profiles. (**a**) CAV-2 elution profile on Fractogel® EMD propyl (S) packed columns. The concentrated CAV-2 feed (70 mL after conditioning to reach appropriate conductivity) was loaded onto a 10 mL HIC column equilibrated in 0.85 M $NH_4(SO_4)_2$ buffer. The virus was eluted by step gradient at 0 M $NH_4(SO_4)_2$. (**b**) CAV-2 elution profile on CIM® DEAE monolithic disks. The semi-purified CAV-2 feed (3 mL after conditioning to reach appropriate conductivity) was loaded onto a 0.34 mL AEX monolithic disk equilibrated in 0.16 M NaCl buffer. The virus was eluted by step gradient at 0.34 M NaCl. The *arrow* indicates CAV-2 elution peak

4. Dilute partially purified CAV-2 samples in buffer A to attain binding conditions (just under 20 mS/cm) (typically 1:4 or 1:4.5). Aliquot starting material for analyses (*see* **Note 21**).
5. When stable baseline is achieved, load the semi-purified virus sample (*see* **Note 22**) and apply a step-wise gradient elution strategy that includes a wash step in binding buffer (22 CV), followed by a virus elution step at 340 mM (44 CV) and a final stringent wash at 1,000 mM NaCl (18 CV). The process is carried out at room temperature at 2 mL/min.
6. Virus particles can be recovered in a peak eluting at 340 mM (~27 mS/cm) (*see* Fig. 3b). Pool virus-containing fractions and aliquot for analyses (*see* **Note 23**).
7. After each run, re-equilibrate the column with binding buffer.

8. Clean the column every three runs. CIM® Q disks can be regenerated by submerging the disk in 1 M NaOH for 2 h. Piston regeneration (the frit regeneration actually) is as follows: 5 min in 1 M NaOH, 5 min in Milli-Q H_2O, 5 min in 100 % PrOH, 5 min in Milli-Q H_2O, and 5 min in a sonicator (ultrasonic bath).
9. Store the disks in storage buffer at 4 °C and purified samples at −80 °C (*see* **Notes 24–26**).

4 Notes

1. The UNICORN system control software associated with the ÄKTA explorer 100 enables on-line monitoring and control of the chromatography process.
2. All buffers should be filtered and degassed prior to chromatography.
3. Fractogel® matrices are very spongy. Do not let these matrices sit in just water as the spacer arms will start clumping together which would seriously hamper the column's ability to capture viral particles. Therefore, even the storage buffer should have at least 0.15 M NaCl in it.
4. Various size CIM® DEAE monolithic columns are commercially available for dealing with larger volumes of semi-purified CAV-2 stocks. This protocol describes the elution strategy that was originally developed using 0.34 mL CIM® DEAE monolithic disks [16].
5. While most CAV-2 particles remain located inside the cell after production, some virus particles may escape to the supernatant fraction. The amount of CAV-2 particles that can be found in the supernatant will vary depending on the specific production procedure employed and should be determined case-by-case. Using our production protocol, 30 % of the active CAV-2 particles are located in the extracellular fraction at the time of vector harvest [16]. The latter can easily be recovered by processing both the extracellular along with the intracellular cell culture fraction.
6. The media used for CAV-2 production should not contain phenol red. Viral stocks are produced in the absence of this pH indicator.
7. Install a pressure gauge inline between the pump outlet and filter inlet to monitor filter inlet pressure at all times. Use appropriate fittings and secure connections. Ensure that the capsule is installed in the correct direction according to the arrow on the capsule label. Check all system tubing to ensure it is not kinked or pinched.

8. The pressure should not exceed the maximum suggested by the manufacturer. At least 3 L of CAV-2 vector stocks can be filtered using 500 cm^2 membrane capsule filters with little backpressure [16].
9. Expected recovery of CAV-2 in the clarified sample using this method is high (84 %, $n=3$) as determined by real-time PCR (qPCR) [16, 19].
10. The hollow fiber cartridge should always be connected in the same direction. Draw arrows to indicate flow direction the first time it is used. Secure all connections. Tubing can be reused. However, it is recommended to replace the tubing connected to the cartridge inlet every two or three runs since it passes through the pump and there is risk for damage. Check all system tubing to ensure it is not kinked or pinched.
11. Alcohol enhances glycerol removal. The inlet (Pi) and outlet pressures (Po) should be low at this point. The permeate pressure (Pp) is ~0 throughout the process.
12. Excessive TMP can cause premature membrane fouling (gel layer effect). Thus, it is important to adjust filtration parameters in early experiments. The optimal TMP is achieved by gradually increasing the TMP by pressing the backpressure valve while monitoring the flow rate. Once the permeate flow rate does not increase by further pressing the valve, the desirable TMP has been reached. Using the Midjet® and Quixstand®, this flow rate was around 1.6 and 11 mL/min, respectively (~60 LMH for both) and the TMP around 1.5 bar or 25 psi, respectively, for our particular CAV-2 feed [16]. Feed volumes of 0.2 L and 1 L could easily be processed using a hollow fiber cartridge with a membrane area of 16 and 110 cm^2, respectively.

    Equations:

$$\text{Transmembrane pressure}(\text{TMP}) = \frac{(\text{Pi} + \text{Po})}{2 - \text{Pp}}$$

$$\text{Recirculation flow rate}(\text{LMH}) = \frac{\text{Flow}(\text{mL}/\text{min}) \times 0.06}{\text{Membrane area}(\text{m}^2)}$$

13. High recovery of CAV-2 in the concentrated and benzonase treated samples is attained (98 %, $n=2$ and 83 %, $n=3$) as determined by qPCR using the Midjet® and Quixstand systems, respectively. A noteworthy removal of contaminating proteins is also accomplished (55 %, $n=2$ and 43 %, $n=3$) [16].
14. The commercial benzonase® stock is typically supplied at a concentration of 250,000 U/mL. To reach a concentration of 10,000 U/mL in the working solution (WS), the original

benzonase® stock is diluted 1/25 in digestion buffer pH 8 just prior to use.

15. Optimal digestion conditions have been previously determined showing over 80 % nucleic acid removal [16].
16. Simultaneous monitoring at 260 and 280 nm will help identify purified virus peaks as purified adenovirus chromatography peaks typically display a 260/280 absorbance ratio between 1.29 and 1.35 [20].
17. It is recommended that samples are not stored at 4 °C for long periods of time between ultrafiltration and chromatography steps, as this practice may result in loss of virus particles upon filtration, presumably due to CAV-2 aggregation.
18. The dynamic capacity of the Fractogel® propyl (S) matrix for CAV-2 particles was determined to be 0.45×10^{12} vg/mL of gel, which typically corresponded to 7 mL of conditioned CAV-2 feed. This capacity is comparable to others described in the literature for adenovirus particles (~0.5 to 5×10^{12} vp/mL) [17].
19. Note that CAV-2 elution from HIC columns typically start as soon as the conductivity drops below 110 mS/cm.
20. Recovery of CAV-2 in the eluted peak is excellent as determined by real-time qPCR (88 %, $n = 9$). In addition, a 3.5-fold concentration of CAV-2 samples in the 10-mL column can be expected.
21. It is recommended that semi-purified samples are subjected AEC directly after HIC, without being stored and with no need for sample microfiltration prior to chromatography. The latter has been associated with loss of virus particles.
22. The dynamic capacity of CIM® DEAE disks for CAV-2 particles was determined to be 0.70×10^{12} vg/mL of monolith, which typically corresponded to 3 mL of semi-purified conditioned CAV-2 feed per disk [16].
23. The majority of virus particles are eluted by step-gradient at 0.34 M NaCl (58 %, $n = 3$) in the first peak eluting at 27 mS/cm (Fig. 3b). A second small peak that could not be separated from the virus elutes during this step at 30 mS/cm. The latter contains only a small fraction of virus particles and can be collected together with the virus increasing the final yield to ~70 % or separately by appropriate sample fractionation. The remaining particles are eluted at 42 mS/cm in the high-stringency wash step at 1 M NaCl (30 %).
24. Due to the weak binding affinity of CAV-2 particles to AEC matrices under the conditions optimized for chromatography, purified samples possess neutral pH and low conductivity. Thus, a subsequent ultrafiltration step to remove salt may not be necessary unless the virus needs to be further concentrated

at the same time. Samples may also be diluted in buffer A before administration to the desired conductivity (i.e., keep in mind that the conductivity of PBS is ~15 mS/cm).

25. The chromatography buffer system used was selected based on the literature to maximize adenovirus particle stability [21].
26. Purity of the final viral products are high as revealed by $OD_{260/280}$ chromatography absorbance ratios (1.3), electron microscopy and electrophoretic analyses [16].

References

1. Sprangers MC, Lakhai W, Koudstaal W, Verhoeven M, Koel BF, Vogels R et al (2003) Quantifying adenovirus-neutralizing antibodies by luciferase transgene detection: addressing preexisting immunity to vaccine and gene therapy vectors. J Clin Microbiol 41:5046–5052
2. Aste-Amezaga M, Bett AJ, Wang F, Casimiro DR, Antonello JM, Patel DK et al (2004) Quantitative adenovirus neutralization assays based on the secreted alkaline phosphatase reporter gene: application in epidemiologic studies and in the design of adenovector vaccines. Hum Gene Ther 15:293–304
3. Bru T, Salinas S, Kremer EJ (2010) An update on canine adenovirus type 2 and its vectors. Viruses 2:2134–2153
4. Soudais C, Laplace-Builhe C, Kissa K, Kremer EJ (2001) Preferential transduction of neurons by canine adenovirus vectors and their efficient retrograde transport in vivo. FASEB J 15:2283–2285
5. Peltekian E, Garcia L, Danos O (2002) Neurotropism and retrograde axonal transport of a canine adenoviral vector: a tool for targeting key structures undergoing neurodegenerative processes. Mol Ther 5:25–32
6. Alcayaga-Miranda F, Cascallo M, Rojas JJ, Pastor J, Alemany R (2010) Osteosarcoma cells as carriers to allow antitumor activity of canine oncolytic adenovirus in the presence of neutralizing antibodies. Cancer Gene Ther 17:792–802
7. Smith BF, Curiel DT, Ternovoi VV, Borovjagin AV, Baker HJ, Cox N et al (2006) Administration of a conditionally replicative oncolytic canine adenovirus in normal dogs. Cancer Biother Radiopharm 21:601–606
8. Hemminki A, Kanerva A, Kremer EJ, Bauerschmitz GJ, Smith BF, Liu B et al (2003) A canine conditionally replicating adenovirus for evaluating oncolytic virotherapy in a syngeneic animal model. Mol Ther 7:163–173
9. Bouet-Cararo C, Contreras V, Fournier A, Jallet C, Guibert JM, Dubois E et al (2011) Canine adenoviruses elicit both humoral and cell-mediated immune responses against rabies following immunisation of sheep. Vaccine 29:1304–1310
10. Henderson H, Jackson F, Bean K, Panasuk B, Niezgoda M, Slate D et al (2009) Oral immunization of raccoons and skunks with a canine adenovirus recombinant rabies vaccine. Vaccine 27:7194–7197
11. Hu R, Zhang S, Fooks AR, Yuan H, Liu Y, Li H et al (2006) Prevention of rabies virus infection in dogs by a recombinant canine adenovirus type-2 encoding the rabies virus glycoprotein. Microbes Infect 8:1090–1097
12. Hu RL, Liu Y, Zhang SF, Zhang F, Fooks AR (2007) Experimental immunization of cats with a recombinant rabies-canine adenovirus vaccine elicits a long-lasting neutralizing antibody response against rabies. Vaccine 25:5301–5307
13. Liu Y, Zhang S, Ma G, Zhang F, Hu R (2008) Efficacy and safety of a live canine adenovirus-vectored rabies virus vaccine in swine. Vaccine 26:5368–5372
14. Dormond E, Perrier M, Kamen A (2009) From the first to the third generation adenoviral vector: what parameters are governing the production yield? Biotechnol Adv 27:133–144
15. Fernandes P, Peixoto C, Santiago VM, Kremer EJ, Coroadinha AS, Alves PM (2013) Bioprocess development for canine adenovirus type 2 vectors. Gene Ther 20(4):353–360
16. Segura MM, Puig M, Monfar M, Chillon M (2012) Chromatography purification of canine adenoviral vectors. Hum Gene Ther Methods 23(3):182–197

17. Altaras NE, Aunins JG, Evans RK, Kamen A, Konz JO, Wolf JJ (2005) Production and formulation of adenovirus vectors. Adv Biochem Eng Biotechnol 99:193–260
18. Segura MM, Kamen AA, Garnier A (2011) Overview of current scalable methods for purification of viral vectors. Methods Mol Biol 737:89–116
19. Segura MM, Monfar M, Puig M, Mennechet F, Ibanes S, Chillon M (2010) A real-time PCR assay for quantification of canine adenoviral vectors. J Virol Methods 163:129–136
20. Tancevski I, Wehinger A, Patsch JR, Ritsch A (2006) In vivo application of adenoviral vectors purified by a Taqman Real Time PCR-supported chromatographic protocol. Int J Biol Macromol 39:77–82
21. Hutchins B (2002) Development of a reference material for characterizing adenovirus vectors. BioProcess J 1:25–29

Chapter 15

Production of High-Capacity Adenovirus Vectors

Florian Kreppel

Abstract

High-capacity adenoviral vectors (HC-Ad), also known as "helper-dependent" (HD-Ad), "gutless", "gutted", or "third-generation" Ad vectors, are devoid of all viral coding sequences and have shown promising potential for a wide variety of different applications—from classic gene therapy to genetic vaccination and tumor treatment. However, compared to first-generation adenoviral vectors their production is more complex and requires specific in-depth knowledge. This chapter delivers a detailed protocol for the successful production of HC-Ad vectors to high titers.

Key words High-capacity adenovirus vectors, Helper-dependent adenovirus vectors, Gutless vectors

1 Introduction

1.1 Features of HC-Ad Vectors

High-capacity adenoviral vectors (HC-Ad), also known as "helper-dependent" (HD-Ad), "gutless", "gutted", or "third-generation" Ad vectors, are devoid of all viral coding sequences. While earlier generations of adenoviral vectors (e.g., ΔE1/ΔE3 Ad vectors) are replication-deficient, they still harbor adenovirus genes which are expressed at a low level after transduction of target cells. This expression of adenovirus genes from the vectors induces very potent CD8+ T cell responses against the transduced cells and, therefore, first- and second-generation adenoviral vectors typically only mediate short-term transgene expression in vivo [1–3].

In contrast, the absence of adenovirus gene expression from HC-Ad vectors allows for long-term transgene expression without chronic toxicity in small and large animal models [4–8]. While being suitable for long-term transgene expression in quiescent or slowly proliferating tissues in vivo, HC-Ad vectors are also characterized by a large cloning capacity. Up to 36 kB of transgenic sequences can be incorporated into the vector genome. This allows for example to clone several expression cassettes and/or large cis-acting elements into one vector genome in order to regulate

Miguel Chillón and Assumpció Bosch (eds.), *Adenovirus: Methods and Protocols*, Methods in Molecular Biology, vol. 1089,
DOI 10.1007/978-1-62703-679-5_15,

transgene expression. As their first-generation counterparts, HC-Ad vectors remain episomally in the nucleus and thus have a very low risk to cause insertional mutagenesis [9].

Beyond classical gene replacement/addition therapy approaches recent results suggest that HC-Ad vectors may also be favorable tools for genetic vaccination, potentially because of the absence of immunocompetition between vector-derived and transgenic antigen epitopes as it occurs with first-generation vectors [10, 11]. Comprehensive overviews over the various features of HC-Ad vectors can be found in ref. [12–14].

1.2 116 Cell-Based HC-Ad Vector Production

Due to the fact that HC-Ad vectors are devoid of all viral coding sequences, their production depends on a helper virus, which delivers the adenoviral gene products required for replication and packaging in trans. This helper virus is a replication-deficient first-generation vector (ΔE1/ΔE3), which can only replicate and deliver the gene products required for HC-Ad vector production in cells that transcomplement *E1*. Various systems have been developed during the last years to allow for preferential packaging of HC-Ad vector genomes in the presence of helper virus genomes [15–19]. The most widely used system is based on a helper virus whose packaging signal Ψ is flanked by cre-recombinase recognition sites (loxP, "floxed packaging signal") [15]. This allows for replication and packaging of HC-Ad vector genomes in the presence of helper virus in cells which transcomplement *E1* and at the same time express cre-recombinase to excise the packaging signal from the helper virus. Obviously, the efficiency with which the packaging signal is excised from the helper virus genomes is a major determinant for the degree of contamination of a HC-Ad vector preparation with helper virus. To our best knowledge the most reliable producer cell line for HC-Ad vectors up to date is the 116 cell line generated by Palmer and Ng [20]. While being based on 293 cells to allow for transcomplementing E1, this cell line is characterized by a high constitutive expression of the Cre-recombinase. In addition, 116 cells can be maintained as adherent monolayers or in suspension culture. The latter significantly facilitates upscaling of vector production.

1.3 HC-Ad Vector Plasmids

The only adenovirus-derived sequences of a HC-Ad vector are the *cis*-acting sequences comprising the inverted terminal repeats (ITRs) and the packaging signal (Ψ), which are required to replicate (ITR) and encapsidate the vector genomes during production. Therefore, the transgene capacity of HC-Ad vectors is about 36 kB which corresponds to the genome size of wild-type adenovirus. Importantly, in order to achieve efficient packaging into stable vector capsids the HC-Ad vector genome should be 28–30 kB in size (including the transgene expression cassette) [21, 22]. If the vector genome is smaller than 27 kB or larger than 37 kB there is a high risk of vector genome rearrangements or low yields. Typical expression cassettes are significantly smaller than 10 kB. Thus, HC-Ad vector genomes are harboring so-called "stuffer" DNA in order to have genome sizes in the preferable range. These stuffer sequences should be derived from noncoding eucaryotic DNA, should be free of repetitive and promoter elements, and should not exhibit homology with the helper virus. A wide variety of different plasmids harboring HC-Ad vector backbones to accommodate differently sized transgene expression cassettes is available.

1.4 Helper Virus

In addition to HC-Ad vector plasmids also a variety of helper virus plasmids is available [11, 20, 23]. In general, the helper virus should—in addition to the E1 deletion and the floxed packaging signal—not harbor any sequences which show homology to the HC-Ad vector.

For HC-Ad vector production only purified and fully characterized helper virus stocks should be used. Helper viruses are produced as first-generation vectors (*see* elsewhere in this book, *see* Chapter 12). We found it mandatory to characterize the genome of the helper virus stocks to the greatest extent possible (including restriction digestion, southern blotting and sequencing) in order to ensure that only helper virus stocks with the correct configuration and sequence of the floxed packaging signal are used.

We have made excellent experience with a helper virus based on pESHV (unpublished), which is an E1-deleted helper virus with a floxed packaging signal derived from Adlc8cluc [15]. In order to generate capsid-modified, retargeted HC-Ad vectors it is mandatory to use an accordingly capsid-modified helper virus [23, 24]. It is sufficient to use this helper virus only during the large-scale production step.

1.4.1 Brief Outline of HC-Ad Vector Production

The production process of HC-Ad vectors is divided into several stages. First, small amounts of infectious HC-Ad vector are rescued from its corresponding plasmid. Second, the small HC-Ad vector amounts are then serially amplified until a certain titer is reached that allows for a medium-scale preparation. Third, this medium-scale preparation is purified, characterized, and used to generate a large-scale preparation. To obtain a high-quality HC-Ad vector preparation within a reasonable time it is mandatory to have a well-organized cell-culture schedule.

2 Materials

2.1 Cultivation 116 Cells

1. 116 producer cells.
2. Minimal essential medium (MEM) for cultivating adherent 116 cells, supplemented with 10 % FCS, 100 U/mL penicillin–streptomycin, 2 mM L-glutamine, and 0.1 mg/mL hygromycin B.

3. MEM eagle, Joklik modification (MEM, Life Technologies/Gibco, Darmstadt, Germany) for cultivating 116 cells in suspension, supplemented with 5 % FCS, 100 U/mL penicillin–streptomycin, 2 mM L-glutamine, and 0.1 mg/mL hygromycin B.
4. Fetal calf serum (FCS).
5. Trypsin–EDTA.
6. Penicillin–Streptomycin–Glutamine.
7. PBS w/o Ca^{2+}, Mg^{2+}.
8. 60-mm and 150-mm cell culture dishes for adherent cells.
9. Spinner flasks, preferably 250-mL and 3 L. Magnetic stirrer.

2.2 HC-Ad Vector Rescue

1. JetPEI (Polyplus, Illkrich, France).
2. HC-Ad vector plasmid.
3. Optional, but recommended: pHC-AdEGFP as control for transfection/amplification.
4. Helper virus.
5. PBS w/o Ca^{2+} and Mg^{2+}.

2.3 HC-Ad Vector Serial Amplifications

1. Optional but recommended: fluorescence microscope with appropriate laser/filter set to excite EGFP (488 nm excitation, 512 nm emission).
2. Helper virus and the corresponding plasmid.
3. QIAamp DNA Mini Kit or equivalent.
4. Glycerol 100 %.
5. Liquid nitrogen and an appropriate container for freeze–thaw.

2.4 HC-Ad Vector Purification

1. Ultracentrifuge, e.g., Beckman XPN-90 with SW41Ti rotor.
2. Ultraclear centrifuge tubes 12.8 mL.
3. HEPES 50 mM pH 8.0.
4. PD-10 disposable desalting columns.
5. Glycerol 100 %.
6. 2 mL syringes and 18-G needles.

3 Methods

3.1 Cultivation of 116 Cells

Careful routine cultivation of 116 cells is mandatory for successful rescue and amplification of HC-Ad vectors. Only healthy cells with a high level of *Cre* expression allow for high-titer HC-Ad vector preparations with low helper virus contamination. Furthermore, a well-organized cultivation schedule guarantees availability of healthy cells when needed for the different amplification steps and thus can save time. 116 cells should be subcultivated when they reach 90 % confluency and it is strongly recommended to never let them become superconfluent. 116 cells are cultivated in 150 mm dishes in supplemented MEM. Use 30 mL of supplemented medium per 150 mm dish. Subcultivation is required at a ratio of 1:3 every 3 days. Pre-warmed media, trypsin, and PBS (37 °C) are recommended for subcultivation.

1. Remove medium by aspiration.
2. Wash cells with 10 mL pre-warmed PBS. Take care to carefully pipette the PBS in order to not detach the cells from the dish.
3. Aspirate PBS.
4. Add 3 mL trypsin to cells. Trypsinize for 1 min at room temperature.
5. Carefully aspirate trypsin.
6. Resuspend the cells in fresh medium and transfer the appropriate volume to new dishes.

Subcultivated 116 cells should be allowed to attach to the new dishes for 48 h prior to infection/transfection for rescue/amplification purposes.

3.2 HC-Ad Vector Rescue

To rescue HC-Ad vectors from their corresponding vector plasmid, the plasmid is linearized, transfected into 116 cells and the cells are coinfected with helper virus. The initial transfection efficiency determines to a significant degree the number of amplifications that are required to obtain a high-titer HC-Ad vector preparation. Optimizing transduction efficiencies is thus mandatory (*see* **Note 1**).

1. For each vector to be rescued seed 116 cells into one 60-mm dish in a way that they reach a confluency of 70–80 % within 48 h.
2. Linearize 15–20 μg of the HC-Ad vector plasmid with the appropriate enzyme in a total volume of 100 μL.
3. Purify the linearized plasmid by phenol/chloroform extraction and ethanol precipitation (*see* **Note 2**).
4. Perform agarose gel electrophoresis and ethidium bromide staining with 100 ng of the linearized and purified plasmid. Only two bands should be visible: a large one comprising the HC-Ad vector genome and a smaller one comprising the bacterial backbone (*see* **Note 3**).
5. 30 min prior to transfection replace the medium of the cells to be transfected with 2 mL of fresh, supplemented MEM.
6. Transfect the cells (one 60-mm dish per vector) with 4 μg of the linearized and purified HC-Ad vector plasmid using jetPEI according to the manufacturer's instructions.
7. One hour after transfection infect the cells with 15–20 MOI of helper virus (*see* **Notes 4** and **5**).

8. One hour after infection with helper virus add 3 mL of fresh, supplemented, pre-warmed medium (the final volume should be 5 mL).
9. 48 h after transfection the cells should show a complete cytopathic effect (CPE) characterized by cell rounding and detachment from the dish (*see* **Note 6**). Harvest the cells in their medium by pipetting up and down or scraping.
10. Spin down cells at 400×*g* for 10 min. Aspirate and discard supernatant. Resuspend the cells in 2 mL PBS. Snap-freeze the cell suspension in liquid nitrogen and store at −80 °C until further use (*see* **Note 7**).

3.3 HC-Ad Vector Serial Amplification

Upon initial transfection only small amounts of HC-Ad vector are produced by the 116 cells (*see* **Note 1**). These small amounts are serially amplified by repeatedly coinfecting fresh 116 cells with purified helper virus and cell lysates containing the HC-Ad vector. As a rule of thumb the HC-Ad titers should increase at least tenfold in each amplification step (*see* **Note 8**).

3.3.1 First HC-Ad Vector Amplification Step

1. Release vector particles from the cells in the suspension obtained in **step 10** of the rescue protocol by three repeated freeze–thaw cycles in liquid nitrogen and a water bath (37 °C). Gently swirl the tubes with the cell suspensions while freezing and thawing.
2. Replace the medium of a 60-mm dish with 90 % confluent 116 cells seeded 2 days before with 2 mL of fresh, supplemented medium.
3. Infect cells with 0.5 mL of the freeze–thaw lysate obtained in **step 1** of the serial amplification protocol (*see* **Note 9**). Coinfect the cells with 15–20 MOI of helper virus (*see* **Note 10**). One hour after infection fill up to 5 mL with supplemented medium (*see* **Note 11**). Supplement the remaining cell lysate with glycerol to a final concentration of 10 % (w/v) prior to storage at −80 °C.
4. 48 h after infection the cells should show a complete cytopathic effect (CPE) characterized by cell rounding and detachment from the dish. Harvest the cells in their medium by pipetting up and down or scraping.
5. Spin down cells at 400×*g* for 10 min. Aspirate and discard supernatant. Resuspend the cells in 2 mL PBS. Snap-freeze the cell suspension in liquid nitrogen and store at −80 °C until further use.

3.3.2 Second HC-Ad Vector Amplification

1. Repeat **steps 1–5**.

3.3.3 Third HC-Ad Vector Amplification

1. Release vector particles from the cells by three repeated freeze–thaw cycles in liquid nitrogen and a water bath (37 °C). Gently swirl the tubes with the cell suspensions while freezing and thawing.
2. Replace the medium of one 150-mm dish with 90 % confluent 116 cells seeded 2 days before with 15 mL of fresh, supplemented medium.
3. Infect cells with 0.5 mL of the freeze–thaw lysate obtained in **step 1**. Coinfect the cells with 15–20 MOI of helper virus. One hour after infection fill up to 30 mL with supplemented medium.
4. 48 h after infection the cells should show a complete cytopathic effect (CPE) characterized by cell rounding and detachment from the dish. Harvest the cells in their medium by pipetting up and down or scraping.
5. Spin down cells at 400 × *g* for 10 min. Aspirate and discard supernatant. Resuspend the cells in 2 mL PBS. Prepare genomic DNA from a 200 μL aliquot of the cell suspension using the QiaAmp DNA Mini Kit (Qiagen, Hilden, Germany) according to the manufacturer's instructions (*see* **Note 12**). Snap-freeze the remaining cell suspension (1.8 mL) in liquid nitrogen and store at −80 °C until further use.

3.3.4 Fourth Vector Amplification

1. Release vector particles from the cells by three repeated freeze–thaw cycles in liquid nitrogen and a water bath (37 °C). Gently swirl the tubes with the cell suspensions while freezing and thawing.
2. Replace the medium of two 150-mm dishes (two dishes per vector) with 90 % confluent 116 cells seeded 2 days before each with 15 mL of fresh, supplemented medium.
3. Infect each dish with 0.5 mL of the freeze–thaw lysate obtained in **step 1**. Coinfect the cells with 15–20 MOI of helper virus. One hour after infection fill up to 30 mL (per dish) with supplemented medium.
4. 48 h after infection the cells should show a complete cytopathic effect (CPE) characterized by cell rounding and detachment from the dish. Harvest the cells in their medium by pipetting up and down or scraping.
5. Spin down cells at 400 × *g* for 10 min. Aspirate and discard supernatant. Combine and resuspend the cells from both dishes in 4 mL PBS. Prepare genomic DNA from a 200 μL aliquot of the cell suspension using the QIAamp DNA Mini Kit according to the manufacturer's instructions. Snap-freeze the remaining cell suspension (3.8 mL) in liquid nitrogen and store at −80 °C until further use.

3.3.5 Decision Point: Intermediate Quality Control

1. Digest the genomic DNA obtained in **step 5** from procedure in Subheadings 3.3.3 and 3.3.4 with an appropriate restriction enzyme (*see* **Note 12**). In parallel, digest 100–200 ng of the linearized HC-Ad vector plasmid obtained in **step 3** of the rescue protocol and 100–200 ng of linearized helper virus plasmid and analyze all samples by agarose gel electrophoresis/ ethidium bromide staining. In the case of successful HC-Ad vector amplification there should be prominent bands visible in the lane loaded with genomic DNA from the fourth amplification step, which correspond to the HC-Ad vector plasmid control. The intensity of these bands should be equal or higher compared to the intensity of the bands which correspond to helper virus genomes. If the intensities of the HC-Ad vector bands is significantly weaker than the intensity of the bands corresponding to helper virus genomes repeat procedure in Subheading 3.3.4 until prominent HC-Ad vector bands become visible in the gel (*see* **Note 13**). If prominent HC-Ad vector bands are visible proceed with the medium-scale preparation of HC-Ad.

3.3.6 Medium-Scale Preparation of HC-Ad

The goal of a medium-scale preparation of the HC-Ad vector at this stage is to obtain the vector in highly purified form and in quantities which are sufficient to perform small-scale functional analysis and, importantly, allow to infect a large number of producer cells in suspension culture in a defined manner to obtain large quantities of the HC-Ad vector. A typical yield of a purified HC-Ad vector from a medium-scale preparation as described in the following is 1E11–1E12 vector particles.

1. Release vector particles from the cells obtained in procedure in Subheading 3.3.4 by three repeated freeze–thaw cycles in liquid nitrogen and a water bath (37 °C). Gently swirl the tubes with the cell suspensions while freezing and thawing.
2. Replace the medium of twelve 150-mm dish with 90 % confluent 116 cells seeded 2 days before with 15 mL of fresh, supplemented medium.
3. Infect cells with 0.3 mL of the freeze–thaw lysate per dish as obtained in **step 1**. Coinfect the cells with 15–20 MOI of helper virus. One hour after infection fill up to 30 mL with supplemented medium per dish.
4. 48 h after infection the cells should show a complete cytopathic effect (CPE) characterized by cell rounding and detachment from the dish. Harvest the cells in their medium by pipetting up and down or scraping.
5. Spin down cells at $400 \times g$ for 10 min. Aspirate and discard supernatant. Resuspend the cells in 4 mL PBS. Snap-freeze the cells in liquid nitrogen and store prior to further use at −80 °C or immediately proceed with purification.

3.4 HC-Ad Vector Purification

Purification of HC-Ad vectors is performed by double CsCl banding and subsequent desalting. The first CsCl banding is based on a discontinuous CsCl gradient, removes >90 % of cellular impurities, and concentrates the vector particles. The second CsCl banding is based on a continuous CsCl gradient. To obtain HC-Ad vector preparations with high purity it is mandatory to have at least one discontinuous CsCl step gradient. Extra high purity can be obtained by performing two subsequent discontinuous CsCl gradients followed by one continuous gradient. The steps given in the following protocol refer to purification of HC-Ad vector particles from twelve 150-mm dishes (approximately 2–3E8 cells). For upscaling *see* **Note 14**.

1. Prepare CsCl solutions with densities of 1.27 g/mL, 1.34 g/mL, and 1.42 g/mL, respectively, in HEPES 50 mM pH 8.0 (*see* **Note 15**).
2. Release the vector particles from the cells obtained in **step 32** of the amplification protocol by three repeated freeze–thaw cycles in liquid nitrogen and a water bath (37 °C).
3. Spin down cell debris by centrifugation at 5,000 × *g* for 10 min at 4 °C and transfer the supernatant containing the released vector particles into a new tube. Centrifuge again at 5,000 × *g* for 10 min at 4 °C (*see* **Note 16**).
4. Pipette 3 mL of the sterile-filtered CsCl solution with a density of 1.42 g/mL into a UltraClear centrifuge tube (14 × 89 mm, 12.8 mL total volume).
5. Carefully overlay with 5 mL of a CsCl solution with a density of 1.27 g/mL.
6. Carefully overlay with the cleared supernatant obtained in **step 3** of the purification protocol.
7. Carefully fill up the remainder of the tubes with PBS.
8. Centrifuge in a Beckman SW41Ti rotor at 32,000 rpm (176,000 × *g*) for 2 h at 4 °C.
9. After centrifugation a vector band should be visible at the interface of the 1.42 g/mL and 1.27 g/mL CsCl solutions. In addition bands comprised of incomplete particles are visible below the interface of the cell lysate and the 1.27 g/mL CsCl solution. Pierce the tube with a 2 mL syringe and 18-G needle below the vector band and slowly retrieve the vector. Take care to not retrieve more than 2 mL.
10. Transfer the vector into a new UltraClear centrifuge tube (14 × 89 mm, 12.8 mL total volume).
11. Fill up with the 1.34 g/mL CsCl solution and mix carefully by pipetting up and down.

12. Centrifuge in a Beckman SW41Ti rotor at 32,000 rpm (176,000 × *g*) for 20 h at 4 °C.
13. After centrifugation usually only one band comprised of the HC-Ad vector is visible. This band should be collected by punctating the tube as described in **step 9** of the purification protocol. If more than one band is visible each band should be collected separately by punctating the tube at the respective position starting with the band with the lowest density (*see* **Note 17**). Take care to collect the individual bands in not more than 2 mL. If more than one band is present in the gradient and collected the following steps of the protocol have to be performed separately for each band.
14. Transfer the collected band to a new reaction tube and fill up with HEPES 50 mM pH 8.0 to a final volume of 2.5 mL. Place the tube on ice.
15. Equilibrate a disposable PD-10 desalting column with 5 × 5 mL HEPES 50 mM pH 8.0 (*see* **Note 18**).
16. Load the HC-Ad vector solution (2.5 mL) onto the PD-10 column. Discard the flow-through.
17. Elute the HC-Ad vector from the PD-10 column with 5 mL HEPES 50 mM pH 8.0. Collect the eluate in 1 mL fractions.
18. Combine fractions 2 and 3 of the eluate, which contain the HC-Ad vector and add 222 μL 100 % glycerol. Mix carefully by pipetting up and down. Remove a 100 μL aliquot from the HC-Ad vector preparation for analysis of the vector DNA and a 20 μL aliquot for a preliminary titer determination by OD measurement.
19. Aliquot the HC-Ad vector and store at −80 °C.

3.5 Characterization of the HC-Ad Vector Genomes

To prove the integrity of the HC-Ad vector genomes it is mandatory to isolate the vector DNA from the particles and perform a restriction digest with at least two different enzymes. As a control 100–200 ng of the linearized vector plasmid from **step 3** of the rescue protocol is digested with the same enzyme in parallel.

1. Isolate vector DNA from the 100 μL aliquot obtained in **step 18** of the purification protocol using the QIAamp DNA Mini Kit according to the manufacturer's instructions.
2. Digest an aliquot of the isolated vector DNA with an appropriate restriction enzyme. As a control digest 100–200 ng of the linearized vector plasmid from **step 3** of the rescue protocol with the same enzyme.
3. Load the samples on an agarose gel, perform ethidium bromide staining and identify the vector bands (*see* **Note 19**).

3.6 Rough Titer Estimation

HC-Ad vector preparations possess a physical particle titer (i.e., the concentration of physical HC-Ad vector particles per volume), an infectious titer (i.e., the number of infectious vector particles able to deliver their genomes into cells per volume), and a contamination with helper virus. All three parameters should be determined to fully characterize HC-Ad vector preparations and a wide variety of DNA-based methods like qPCR and slot-blotting are available. For the generation of high-titer HC-Ad vector stocks from purified HC-Ad vector it is sufficient to roughly determine the physical particle titer by OD260 (*see* **Note 20**).

1. Add 79 μL of 50 mM HEPES pH 8.0 to the 20 μL aliquot of purified HC-Ad vector as obtained in **step 18** of the purification protocol.
2. Add 1 μL of a 10 % solution of SDS in water (w/v).
3. Mix and incubate at 56 °C for 10 min.
4. Allow the sample to cool down to room temperature. Do *not* put the sample on ice.
5. Measure OD at 260 nm.
6. Calculate the physical particle titer: OD260 × 5 × 1.1E12 = vp/mL

3.7 Large-Scale Preparation of HC-Ad Vectors

Large-scale preparations of HC-Ad vectors are most conveniently done with 116 cells in spinner culture. While 116 cells in spinner culture can be infected with crude cell lysates obtained from serial vector amplifications it is recommended to use purified medium-scale preparations obtained by the protocol described in Subheading 3.4. The reason for that is that it is difficult to evaluate the CPE of cells in spinner culture and using defined amounts of a purified and titrated medium-scale vector preparation helps to circumvent this pitfall.

1. Allow ten 150-mm dishes of adherent 116 cells to grow to 90 % confluency.
2. Aspirate the medium from the cells and add 3 mL of pre-warmed trypsin per dish.
3. Trypsinize cells for 1 min at room temperature and aspirate trypsin.
4. Resuspend the cells of the ten dishes in 300 mL of fresh, pre-warmed, MEM-Joklik supplemented with 5 % FCS, 0.1 mg/mL hygromycin, 100 U/mL penicillin–streptomycin, and 2 mM glutamine. Transfer the cell suspension to a 3 L spinner flask.
5. Incubate at 37 °C and stir cells at 60 rpm.
6. On the next day take an aliquot of the suspension, microscopically examine and count the cells. Add 0.5 L of fresh medium supplemented as described in **step 4**.

7. On the next day take an aliquot of the suspension, microscopically examine and count the cells. Add 0.5 L of fresh medium supplemented as described in **step 4**.
8. On the next day take an aliquot of the suspension, microscopically examine and count the cells. Add 1.0 L of fresh medium supplemented as described in **step 4**.
9. On the next day take an aliquot of the cell suspension and count the cells. The cell density should be between 2 and 4E5 cells/mL at the time of coinfection with helper virus and HC-Ad vector. In addition, the cells should not have formed clumps.
10. When the cells reach the required density, harvest the cells by centrifugation at 400 × *g* for 10 min.
11. Resuspend the cells in 100 mL of the spent medium and transfer them to a fresh 250-mL spinner flask.
12. Infect cells with 200–300 physical particles of a purified and characterized HC-Ad vector preparation and coinfect with 15–20 MOI helper virus.
13. Incubate the cells for 2 h at 37 °C while stirring at 60 rpm.
14. Transfer the cells including the medium to a fresh 3 L spinner flask and fill up to 3 L with fresh medium.
15. Incubate for 48 h while stirring at 60 rpm.
16. Harvest the cells by centrifugation at 400 × *g* and resuspend in 4 mL PBS per 2E8 cells.
17. Continue with the purification protocol described in Subheading 3.4 (*see* **Notes 21** and **22**).

3.8 Handling, Storage, and Shipping

1. HC-Ad vectors can be stored long-term (>4 years) at −80 °C when supplemented with 10 % glycerol as cryoprotectant.
2. Avoid repeated freezing and thawing of HC-Ad vector preparations by preparing appropriately sized aliquots before the first freezing.
3. Avoid prolonged storage at room temperature.
4. Avoid storage in glass containers. Depending on the nature of the glass the vector may irreversibly adhere to the glass surface.
5. Avoid vortexing or other means of vigorous mixing.
6. When shipping on dry ice prevent CO_2 from entering the packaging of the vector. It is best to place the vector tubes wrapped into parafilm into several layers of CO_2-tight plastic bags, which are vacuum sealed.

4 Notes

1. It is recommended to include a transfection control with a linearized HC-Ad vector plasmid that expresses a reporter gene like *EGFP* or *ß-gal*. *EGFP* is preferable since it allows for determining transfection efficiencies using a conventional fluorescence microscope and does not require staining procedures. The transfection efficiency should be >60 %. Attempts to rescue and amplify HC-Ad vectors from their corresponding plasmid at low transfection rates (<40 %) frequently result in a large number of serial amplifications and, importantly, high helper virus contaminations and/or vector genome rearrangements. High transfection rates in the rescue step are required, because only few copies of the transfected plasmid will undergo replication and be packaged into capsids for reasons that are not yet fully understood.
2. Purification of the linearized HC-Ad vector plasmid can significantly increase transfection efficiencies and phenol/chloroform extraction typically results in higher yields and higher quality DNA compared to commercially available spin columns.
3. When performing agarose gel electrophoresis/ethidium bromide staining with the linearized HC-Ad vector plasmid, carefully check if there is DNA remaining in the pocket of the gel. DNA remaining in the loading pocket indicates that the purity of the HC-Ad vector DNA is not sufficient for transfection. In that case phenol/chloroform extraction should be repeated.
4. One hour incubation of the polyplexes with the cells allows for adherence of the polyplexes to the cell surface. The infection with helper virus at that time will significantly increase transfection rates since the concomitant uptake of polyplex and helper virus increases the efficiency of endosomal escape of the polyplexes and reprograms the cytoskeleton for transport to the nuclear pores.
5. The infectious titer of the helper virus should not be determined by a plaque assay, since this method is difficult to reproduce. Plaque formation requires a series of events after transduction of a cell and is thus not representative for the infectious titer of the helper virus. Consistently pfu titers are typically five- to tenfold below the infectious titer determined by DNA-based methods. We recommend to determine the infectious titer either by qPCR or by a DNA-based slot-blot procedure [25, 26].
6. The optimal CPE is characterized by >90 % of the cells being rounded and detached from the dish in clusters. If more than 10 % of the cells are not rounded and still attached to the dish

the helper virus amount was too low and the transfection/coinfection should be repeated with an adjusted amount of helper virus. If all cells are rounded and detached but the majority is present as single cells (instead of clusters) the helper virus amount was too high and the transfection/amplification should be repeated with an adjusted amount of helper virus. For all steps of rescue and amplification the optimal CPE should occur within 40–48 h. If it occurs earlier or later the amount of helper virus should be adjusted and the transfection/amplification step should be repeated with an adjusted amount of helper virus. For researches who are not or only marginally familiar with high-titer production of first-generation adenovirus vectors it is strongly recommended to learn to microscopically evaluate the completeness of a CPE by infecting a series of dishes with 116 cells with increasing amounts of helper virus and observing the (sometimes subtle) morphological changes over time. In addition, for trouble shooting it is advised to take photographs of the different stages of CPE formation during HC-Ad vector amplification. Please note that while the microscopic evaluation of the CPE is an indispensable tool for successful rescue and amplification of HC-Ad vectors, the appearance of the CPE may—in addition to the amounts of helper virus used—be influenced by the transgene expressed from the HC-Ad vector to be amplified.

7. Cells containing HC-Ad vector from the different rescue/amplification steps can be snap-frozen and stored at −80 °C without cryoprotectant until further use. It is not recommended to store lysates of the cells (i.e., lysates obtained by repeated freeze–thaw cycles as in **steps 11**, **17**, **22**) for prolonged periods at −80 °C without the addition of a cryoprotectant (10 % w/v glycerol).
8. We recommend to amplify the HC-AdEGFP vector used as a transfection control (*see* **Note 1**) in parallel to the actual HC-Ad vector to be produced, because for the HC-AdEGFP vector the process of amplification can easily be visualized using a fluorescence microscope. Monitoring the amplification process gives valuable hints on how to optimize the different amplification steps. However, it should be noted that vectors with different transgenes (e.g., toxic transgenes or transgenes that interfere with adenovirus replication) can behave differently during the amplification process.
9. There is no need to clear the cell lysates by centrifugation after freeze–thaw. Clearing the cell lysates may result in spinning down vector which is attached to membrane fragments. Instead of clearing the lysates take care to also use the cell debris in the infection step. This holds true for all amplification steps.

10. In the vast majority of cases coinfection with cell lysate and helper virus at the same time allows to amplify the HC-Ad vector of choice. However, we found for a HC-Ad vector expressing the *EBNA-1* gene from Epstein-Barr virus, which is suspected to interfere with Ad vector replication, that only infecting the cells with helper virus 6 h prior to infection with the cell lysate allowed for amplification of HC-AdEBNA-1 to high titers [9]. If you suspect your transgene to interfere with Ad replication it may be worth to try this or a similar infection schedule throughout the whole amplification/production process.
11. Check the dishes infected with lysates obtained from transfection/coinfection of pHC-AdEGFP for EGFP expression 24 h after infection. There should at least be 5–10 % EGFP positive cells in your dish. If the number of EGFP positive cells is significantly lower your rescue step was inefficient and should be repeated. Do not be mislead by this low number of EGFP positive cells (compared to initial transfection). This reflects the low efficiency with which HC-Ad vectors can be rescued from linearized plasmids. Importantly, in all subsequent amplification steps there should be an at least tenfold increase in titer.
12. During HC-Ad vector production large quantities of HC-Ad vector and helper virus genomes are produced. Note that the helper virus genomes are replicated but not packaged into virions due to excision of the packaging signal by the cre-recombinase. Thus, the presence of helper virus genomes in the genomic DNA isolates is not indicative for a helper virus contamination. However, the isolation of genomic DNA from producer cells at the time point of harvesting, digestion of the DNA with an appropriate restriction enzyme, and analysis by agarose gel electrophoresis/ethidium bromide staining allows to estimate the success of vector amplification.
13. The number of amplification steps required for successful HC-Ad vector generation ranges from 4 to 7 and is influenced by a wide variety of parameters including the HC-Ad vector genome size (ideally between 30 and 35 kb), the toxicity of the transgene expressed from the vector, and the presence of repetitive DNA elements in the HC-Ad vector (which should be avoided). However, if more than seven serial amplification steps are required it is likely that the final vector preparation will contain a high helper virus contamination and/or rearranged HC-Ad vector genomes. In that case optimizing each single amplification step by varying the amounts of helper virus and cell lysate added is strongly recommended.
14. Typical yields of HC-Ad vector allow to purify the vector produced from 2 to 4E8 cells (lysate in 4 mL) in one Beckman

UltraClear tube (14 × 89 mm, 12.8 mL total volume). If more cells are used (for example, in a large-scale preparation) accordingly more UltraClear tubes are required.

15. CsCl solutions should be prepared in an appropriate buffer system and sterile filtered. We have excellent experience using 50 mM HEPES as buffer system. In that case, the storage bottles should be wrapped into aluminum to protect from light. Alternative buffer systems, e.g., 100 mM Tris may be used. The solutions can be stored for prolonged time at room temperature, but each solution should be mixed prior to use. It is recommended to verify the density of each solution by weighing 1 mL immediately prior to use. To prepare 100 mL of a CsCl solution with a density of 1.27 g/mL dissolve 36.94 g CsCl in a final volume of 100 mL 50 mM HEPES pH 8.0. To prepare 100 mL of a CsCl solution with a density of 1.34 g/mL dissolve 47.2 g CsCl in a final volume of 100 mL 50 mM HEPES pH 8.0. To prepare 100 mL of a CsCl solution with a density of 1.42 g/mL dissolve 56.78 g in a final volume of 100 mL 50 mM HEPES pH 8.0.
16. Clearing the cell lysates prior to loading onto the CsCl step gradients is mandatory. If the cell lysates are not cleared by centrifugation before loading onto the gradient, the gradient will be disturbed and the purity of the vector preparation will be low. It is recommended to keep the cell debris (supplemented with 10 % glycerol and stored at −80 °C), which still contains significant amounts of HC-Ad vector since it can be used for reinfecting fresh producer cells for another round of vector production.
17. The appearance of more than one band in the continuous CsCl gradient can have several causes. First, one additional band can be comprised of helper virus. Second, additional bands can be comprised of rearranged vector genomes. In any case the bands should be collected separately, DNA should be prepared from the vector particles and analyzed by restriction digestion and agarose gel electrophoresis and/or Southern blotting. It is not recommended to use a HC-Ad vector preparation for which several bands occurred in the continuous gradient for the reinfection of fresh producer cells.
18. Alternative protocols use dialysis as the preferred method for desalting HC-Ad vector preparations. Compared to dialysis the disposable PD-10 desalting columns are time saving.
19. DNA isolated from Ad vector particles can be difficult to digest. It is recommended to use robust enzymes and perform overnight digestions. The linearized plasmid control will

typically yield one additional band, which is not present in the digestion of the HC-Ad vector DNA and comprises the plasmid backbone. All other bands should be visible in both the plasmid control and the HC-Ad vector genome. If the digestion of the HC-Ad vector genome reveals additional bands, which are not present in the plasmid control this is a hint for either a high helper virus contamination or rearranged vector genomes. In that case a more sensitive Southern blot analysis is recommended.

20. Measuring the OD at 260 nm after rupture of the particles by SDS is the easiest way to determine the physical particle titer of a HC-Ad vector preparation. It should be noted that the accuracy of this method very strongly depends on the purity of the vector preparation. Any contamination with nucleic acids or protein will make the titer appear higher than it is. We therefore recommend to routinely analyze the purity of HC-Ad vector preparations by SDS-PAGE and subsequent silver staining of the gels. Loading of 1E10 particles is sufficient to visualize hexon (100 kDa), penton base, IIIa, and fiber (60–65 kDa) capsid proteins. Any additional bands in the silver-stained gel, which cannot be attributed to vector capsid/core proteins indicate impurities. In that case the vector should be re-banded with a CsCl step gradient. To obtain reliable titers by OD260 it is also important to not exceed the final SDS concentration of 0.1 % because at higher concentrations SDS starts to form micelles, which influence the absorbance.
21. Large-scale HC-Ad vector preparations can be used to coinfect fresh producer cells with the HC-Ad vector and helper virus to obtain another large-scale preparation. However, due to the fact that a contamination with mutated helper virus or rearranged HC-Ad vector can occur and may overgrow the actual HC-Ad vector. Therefore, it is advised to use one fully characterized HC-Ad vector preparation as a master stock instead of generating grand- and grand-grand-daughter preparations over several generations.
22. The contamination of HC-Ad vector preparations with helper virus can vary, but should be at least below 1 % and preferably below 0.1 %. If helper virus contaminations are above 1 % it is advised to adjust both the amount of helper virus and the amount of lysate added during the serial amplification steps and the final production step. Again, it should be noted that only helper virus preparations whose sequence of the floxed packaging signal is fully confirmed should be used.

References

1. Yang Y, Nunes FA, Berencsi K, Furth EE, Gönczöl E, Wilson JM (1994) Cellular immunity to viral antigens limits E1-deleted adenoviruses for gene therapy. Proc Natl Acad Sci USA 91:4407–4411
2. Engelhardt JF, Litzky L, Wilson JM (1994) Prolonged transgene expression in cotton rat lung with recombinant adenoviruses defective in E2a. Hum Gene Ther 5:1217–1229
3. Yang Y, Ertl HC, Wilson JM (1994) MHC class I-restricted cytotoxic T lymphocytes to viral antigens destroy hepatocytes in mice infected with E1-deleted recombinant adenoviruses. Immunity 1:433–442
4. Schiedner G, Morral N, Parks RJ, Wu Y, Koopmans SC, Langston C et al (1998) Genomic DNA transfer with a high-capacity adenovirus vector results in improved in vivo gene expression and decreased toxicity. Nat Genet 18:180–183
5. Ehrhardt A, Xu H, Dillow AM, Bellinger DA, Nichols TC, Kay MA (2003) A gene-deleted adenoviral vector results in phenotypic correction of canine hemophilia B without liver toxicity or thrombocytopenia. Blood 102: 2403–2411
6. Brunetti-Pierri N, Palmer DJ, Mane V, Finegold M, Beaudet al, Ng P (2005) Increased hepatic transduction with reduced systemic dissemination and proinflammatory cytokines following hydrodynamic injection of helper-dependent adenoviral vectors. Mol Ther 12:99–106
7. Brunetti-Pierri N, Ng T, Iannitti DA, Palmer DJ, Beaudet al, Finegold MJ et al (2006) Improved hepatic transduction, reduced systemic vector dissemination, and long-term transgene expression by delivering helper-dependent adenoviral vectors into the surgically isolated liver of nonhuman primates. Hum Gene Ther 17:391–404
8. Brunetti-Pierri N, Liou A, Patel P, Palmer D, Grove N, Finegold M et al (2012) Balloon catheter delivery of helper-dependent adenoviral vector results in sustained, therapeutic hFIX expression in rhesus macaques. Mol Ther 20:1863–1870
9. Kreppel F, Kochanek S (2004) Long-term transgene expression in proliferating cells mediated by episomally maintained high-capacity adenovirus vectors. J Virol 78:9–22
10. Schirmbeck R, Reimann J, Kochanek S, Kreppel F (2008) The immunogenicity of adenovirus vectors limits the multispecificity of CD8 T-cell responses to vector-encoded transgenic antigens. Mol Ther 16:1609–1616
11. Kron MW, Engler T, Schmidt E, Schirmbeck R, Kochanek S, Kreppel F (2011) High-capacity adenoviral vectors circumvent the limitations of ΔE1 and ΔE1/ΔE3 adenovirus vectors to induce multispecific transgene product-directed CD8 T-cell responses. J Gene Med 13:648–657
12. Kochanek S (1999) High-capacity adenoviral vectors for gene transfer and somatic gene therapy. Hum Gene Ther 10:2451–2459
13. Brunetti-Pierri N, Ng P (2011) Helper-dependent adenoviral vectors for liver-directed gene therapy. Hum Mol Genet 20:R7–R13
14. Segura MM, Alba R, Bosch A, Chillón M (2008) Advances in helper-dependent adenoviral vector research. Curr Gene Ther 8:222–235
15. Parks RJ, Chen L, Anton M, Sankar U, Rudnicki MA, Graham FL (1996) A helper-dependent adenovirus vector system: removal of helper virus by Cre-mediated excision of the viral packaging signal. Proc Natl Acad Sci USA 93:13565–13570
16. Ng P, Beauchamp C, Evelegh C, Parks R, Graham FL (2001) Development of a FLP/frt system for generating helper-dependent adenoviral vectors. Mol Ther 3:809–815
17. Umaña P, Gerdes CA, Stone D, Davis JR, Ward D, Castro MG et al (2001) Efficient FLPe recombinase enables scalable production of helper-dependent adenoviral vectors with negligible helper-virus contamination. Nat Biotechnol 19:582–585
18. Dormond E, Meneses-Acosta A, Jacob D, Durocher Y, Gilbert R, Perrier M et al (2009) An efficient and scalable process for helper-dependent adenoviral vector production using polyethylenimine-adenofection. Biotechnol Bioeng 102:800–810
19. Alba R, Hearing P, Bosch A, Chillon M (2007) Differential amplification of adenovirus vectors by flanking the packaging signal with attB/attP-PhiC31 sequences: implications for helper-dependent adenovirus production. Virology 367:51–58
20. Palmer D, Ng P (2003) Improved system for helper-dependent adenoviral vector production. Mol Ther 8:846–852
21. Parks RJ, Bramson JL, Wan Y, Addison CL, Graham FL (1999) Effects of stuffer DNA on transgene expression from helper-dependent adenovirus vectors. J Virol 73:8027–8034

22. Kennedy MA, Parks RJ (2009) Adenovirus virion stability and the viral genome: size matters. Mol Ther 17:1664–1666
23. Prill J-M, Espenlaub S, Samen U, Engler T, Schmidt E, Vetrini F et al (2011) Modifications of adenovirus hexon allow for either hepatocyte detargeting or targeting with potential evasion from Kupffer cells. Mol Ther 19:83–92
24. Guse K, Suzuki M, Sule G, Bertin TK, Tyynismaa H, Ahola-Erkkilä S et al (2012) Capsid-modified adenoviral vectors for improved muscle-directed gene therapy. Hum Gene Ther 23:1065–1070
25. Kreppel F, Biermann V, Kochanek S, Schiedner G (2002) A DNA-based method to assay total and infectious particle contents and helper virus contamination in high-capacity adenoviral vector preparations. Hum Gene Ther 13:1151–1156
26. Palmer DJ, Ng P (2004) Physical and infectious titers of helper-dependent adenoviral vectors: a method of direct comparison to the adenovirus reference material. Mol Ther 10:792–798

Chapter 16

Production of Chimeric Adenovirus

Marta Miralles, Marc Garcia, Marcos Tejero, Assumpció Bosch, and Miguel Chillón

Abstract

The use of chimeric pseudotyped vectors is a common way to modify the adenoviral tropism by replacing the fiber protein. In this chapter the procedure to generate a chimeric adenovirus pre-stock from a plasmid containing the adenoviral genome is described. Also, the chimeric adenovirus replicative cycle to increase the yield in further productions is determined. Finally, two different protocols, in culture plates and in suspension cultures, to produce the virus at large scale are also detailed.

Key words Adenoviral vector, Chimeric vectors, Adenovirus production

1 Introduction

Although more than 50 serotypes of HAdVs have been described [1], group C (HAdV-2 and HAdV-5) is still the most studied group [2]. Thus, vectors based on human adenovirus serotype 5 (HAdV-5) have been widely employed as vehicles for many strategies because they have several advantages over other gene delivery systems. These vectors can be amplified to very high titers, which is critical for in vivo assays and clinical applications [3]. Moreover, HAdV-5 vectors are relatively safe because HAdV genome does not integrate into the host genome [4, 5]. For these reasons, HAdV-5 and chimeric HAdV-5 derived vectors have been preferably used among other serotypes.

The term "chimeric virus" is generally used for a recombinant virus generated by the combination of two different viral genomes. Thus, the new chimeric virus may display a combination of the biological properties of both parent viruses. Usually, chimeric adenoviruses have been generated by fiber replacement, involving the HAdV-5 backbone and the fiber protein of a different serotype to thus modify the transduction efficiency and/or the transduction

Miguel Chillón and Assumpció Bosch (eds.), *Adenovirus: Methods and Protocols*, Methods in Molecular Biology, vol. 1089, DOI 10.1007/978-1-62703-679-5_16, © Springer Science+Business Media, LLC 2014

specificity [6, 7]. Combination of other components may also allow to evade the host immune response [8] or to induce different gene expression or gene regulation profiles.

As chimeric adenoviruses maintain most of their original genes, their replication and assembling mechanisms should be maintained. Therefore, chimeric HAdV-5-derived vectors must to be produced in permissible HEK-293 cells. However, since chimerism can alter infectivity, efficiency, virion intracellular trafficking, viral genome replication, and viral assembly, the standard production protocols have to be adjusted for every new chimeric vector. Indeed, as other adenoviral-derived vectors, chimeric adenoviruses are genetically modified organisms (GMO) and have to be produced on a Biosafety Level 2 laboratory.

2 Materials

2.1 Generation of Viral Pre-stock

1. Dulbecco's Modified Eagle's Medium (DMEM) supplemented with 10 % or 2 % Fetal Bovine Serum (FBS).
2. HEK-293 cells.
3. *Pac*I restriction enzyme.
4. 150 mM NaCl.
5. Polyethylenimine (PEI).

2.2 Purification of Viral Pre-stock

1. Ultracentrifuge: Beckman Coulter Optima L90K o L100XP with rotors SW32 and SW40Ti.
2. Polyallomer centrifuge tubes for SW32 and SW40 rotor.
3. CsCl solutions: 1.4 g/mL, 1.34 g/mL, and 1.25 g/mL in 1× PBS.
4. 18G needles, 2 mL syringes, pipette-aid, 10 mL pipettes.
5. PD-10 columns Sephadex G-25.
6. 1× PBS Ca^{2+}/Mg^{2+}.
7. Glycerol, anhydrid.

2.3 Tittering and Analysis of Adenoviral Replicative Cycle

1. A viral pre-stock (fully characterized). It is recommended to use vectors carrying a reporter gene.
2. HEK-293 cells.
3. Growing medium: DMEM supplemented with 10 % FBS and 1 % Penicillin–Streptomycin.
4. Infection medium: DMEM supplemented with 2 % FBS and 1 % Penicillin–Streptomycin.
5. 24-well and 96-well plates.

2.4 Chimeric Adenoviral Production from a Pre-stock

1. 150-mm petri dishes (20 plates).
2. HEK-293 cells.
3. Growing and infection medium as described in Subheading 2.3.
4. CsCl solutions: 1.4 g/mL, 1.34 g/mL, and 1.25 g/mL in 1× D-PBS.
5. Ultracentrifuge and rotors: Beckman Coulter Optima L90K or L100XP. Swinging bucket rotors SW32 and SW40ti.
6. Centrifuge tubes for SW32 and SW40ti rotors.
7. 2 and 10 mL syringes, and 18G needles.
8. Glycerol, anhydride.
9. PD-10 columns Sephadex G-25.
10. 1× D-PBS Ca^{2+}/Mg^{2+}.
11. 96-Well plates.
12. Optional: Polybrene.

2.5 Production in Suspension 211BS Cell Cultures

1. Polycarbonate shake flasks, 125 mL and 1 L.
2. Infection Medium for suspension cells: Freestyle serum-free medium (12338-018, Invitrogen) supplemented with 100 U/mL Penicillin, 0.1 mg/mL Streptomycin, and 1 % Pluronics.
3. 211BS cells [9].
4. Polybrene.
5. Growing medium for suspension cells: SFMII medium (11686-029, GIBCO) supplemented with 4 mM Glutamine, 100 U/mL Penicillin, 0.1 mg/mL Streptomycin, and 1 % Pluronics.
6. Midjet System (GE Healthcare).

3 Methods

3.1 Generation of Viral Pre-stock

It is highly recommended to perform a small pre-stock to generate enough viruses for the following amplification steps. As general trend, for each amplification step it is recommended to wait for cytophatic effect (CPE) to harvest the infected cells. Finally, viral infectivity, viral productivity, and viral cycle must be analyzed to characterize the viral production.

1. Digest 100 μg of the recombinant adenoviral plasmid with 30 Units of *Pac*I restriction enzyme in a total volume of 100 μL to remove the plasmid backbone of bacterial origin sequences. Digest for 4 h at 37 °C (*see* **Note 1**).
2. Add another 30 Units of *Pac*I to guarantee a complete digestion. Digest 4 h at 37 °C.
3. Precipitate the digested DNA with 2 volumes of ethanol and resuspend in 50 μL of sterile ddH_2O.

4. Transfect 10^6 HEK-293 cells plated at 60–70 % of confluency with 6 μg of *Pac*I-digested plasmid (*see* **Note 2**).
5. After 3 days, scrape the cells. Freeze/thaw three times the crude lysate to release adenoviral particles from the cells.
6. Centrifuge at 4,500×*g* for 5 min. Save the supernatant and discard the pellet to remove the cell debris.
7. Seed one 10-cm plate with HEK-293 cells at 60–70 % of confluency. Add 3 mL of DMEM 2 % FBS on the cells as well as all the supernatant from the previous step (*see* **Note 3**).
8. Check the cells during a period of 5–20 days until a 50 % of CPE is observed. Harvest medium and cells and freeze/thaw three times to release adenoviral particles from the cells.
9. Centrifuge at 4,500×*g* for 5 min. Save the supernatant and discard the pellet to remove the cell debris.
10. Seed one 150-mm plate with HEK-293 cells at 60–70 % of confluence (15×10^6 HEK-293 cells). Use DMEM 2 % FBS as medium.
11. Add the supernatant from **step 9** to the cells.
12. Follow the cytopathic effect during the first 2–4 days. Unlike the previous amplification step, the CPE should appear in a period of 30–48 h post-infection. Nevertheless, if vector's productivity or infectivity is poor, check the cells during a period of 5–20 days until a 50 % of CPE is observed. Harvest medium and cells, and freeze/thaw three times to release the adenovirus from the cells.
13. Centrifuge at 4,500×*g* for 5 min. Keep the supernatant and discard the pellet to remove the cell debris.
14. Seed 10×150-mm cell plate with HEK-293 cells at 60–70 % of confluency. Use infection medium.
15. Distribute the supernatant from **step 13** among the ten plates.
16. Check the cells during a period of 2–4 days until a 50 % of CPE is observed as previously describe. Harvest medium and cells.
17. Harvest the cells and resuspend the cell pellet in 18 mL of supernatant. Freeze/thaw three times to release the adenovirus from the cells.
18. Titrate the viral pre-stock (*see* Subheading 3.3).

3.2 Purification of Viral Pre-stock

1. Add 10 mL of 1.25 g/mL CsCl solution in two SW32 centrifuge tubes.
2. Add carefully 10 mL of 1.40 g/mL CsCl solution at the bottom of each SW32 centrifuge tubes to generate two density phases. Take care to not disturb the gradient when removing the pipette.

3. Add the total volume of the cell pellet (18 mL) in one SW32 tube and the precipitated solution from the supernatant in another SW32 tube on the top of the gradient and equilibrate tubes with 1×-PBS.
4. Centrifuge for 1 h 42 min at 125,500×*g* at 18 °C in a Beckman SW32 rotor with maximum brake.
5. Vector appears as an opaque band at the interface between 1.25 g/mL and 1.4 g/mL CsCl. Collect the band by piercing the SW32 tube about 1 cm below the band using a 2 mL syringe loaded with a 18G needle.
6. Add 5 mL of 1.32 g/mL CsCl solution in one SW40 centrifuge tube and fill a second SW40 tube with 5 mL of the 1.32 g/mL CsCl solution.
7. Add the recovered vector band from the previous step on the top of the 1.32 g/mL CsCl solution. Equilibrate with 1×-PBS.
8. Centrifuge for 22 h at 155,000×*g* 18 °C in a Beckman SW40 rotor with maximum brake.
9. Vector appears as opaque band in the middle of the tube. Collect the band as above in less than 2 mL of volume.
10. For each band, prepare ten eppendorf tubes to collect the viral fractions.
11. Pass through PD-10 column, 25 mL of 1× PBS Ca^{2+}/Mg^{2+} to remove the preservation solution and equilibrate the column. Discard the flow-through.
12. Add the total volume of the collected band from **step 9**. Discard the flow-through.
13. Add 500 μL of 1× PBS Ca^{2+}/Mg^{2+} to the PD-10 column and collect the flow-through in an eppendorf tube.
14. Repeat the previous step until 10 fractions per band are collected.
15. Add glycerol to a final concentration of 10 % to each fraction (approx. 55 μL per fraction).
16. Store the purified viral fractions at −80 °C.

3.3 Measurement of Viral Infectivity

The infectivity of a viral particle is the capability to enter into a cell and express its genome. The infectivity depends on multiple factors such as the cell line, its passage, or the properties of the viral particle by itself, and it can be altered in a new chimeric adenovirus. Usually, the infectivity of the Human Adenovirus type 5 in HEK-293 cells ranges between 1/10 and 1/100 (IU/vg; Infection Units per viral genomes). Although the viral backbone may be almost identical, the infectivity of chimeric adenoviruses may differ significantly from HAdV-5. Thus, HAdV-5/40 has a poor IU/vg ratio (between 1/100 and 1/600) [9].

3.3.1 First Titer of the Viral Fractions

1. Seed HEK-293 cells in a 96-well plate at 60–70 % of confluency in a total volume of 100 μL/well of infection medium.
2. Twenty-four hours later, prepare serial dilutions of the ten viral fractions from **step 16** in Subheading 3.2. Prepare each dilution in a final volume of 500 μL. Use DMEM supplemented with 2 % FBS for the dilutions. Typically, start with a $1/10^6$ dilution and make 1/10 serial dilutions. Perform 6 dilutions ($n=2$ per dilution).
3. Carefully, remove the media from the HEK-293 cells and infect with 100 μL of each dilution (*see* **Note 4**).
4. Two days after infection, count the number of infected cells using your reporter gene (i.e., GFP or β-galactosidase) or by immunocytochemistry using the anti-hexon-antibody-based system as described in Chapter 12. For calculations, consider only dilutions for wells with less than 10 % of positive cells.
5. Calculate the titer expressed in Infection Units/mL (IU/mL) using the following equation:

$$\text{Titer}\left(\text{IU}/\text{mL}\right)=\frac{\text{Positive cells}\times\text{Dilution factor}\times\text{infection volume}}{10^3} \quad (1)$$

6. Pool the three or four most concentrated fractions (usually fractions 4–7) and aliquot the viral pre-stock (*see* **Note 5**) in 0.5 mL tubes and store at −80 °C as quickly as possible.

3.3.2 Final Titer of the Viral Pre-stock

1. Repeat **steps 1–5** from procedure in Subheading 3.3.1, but start with a $1/10^9$ viral dilution and make ½ serial dilutions as previously explain. Make 12 dilutions.
2. Calculate the titer applying Eq. 1. When working with a fluorescent reporter gene and using cells in which the vector is replicative, select for calculations the most diluted condition showing positive cells 5–7 days after the infection and consider the number of positive cells as "1" (*see* **Note 6**).
3. Prepare a 1/20, 1/10, and 1/5 viral dilutions in a total volume of 100 μL. Use 1× PBS Ca^{2+}/Mg^{2+} plus 0.1 % SDS for the dilutions (*see* **Note 7**).
4. Incubate at 56 °C for 10 min to disrupt viral capsids and release viral genomes.
5. Centrifuge the tubes at 4,500 × *g* for 5 min.
6. Measure the optical density of the supernatants at 260 nm (OD_{260}) (*see* **Note 8**).
7. Calculate the titer expressed in viral genomes/mL (vg/mL) using the following equation:

$$\text{Titer}\left(\text{vg}/\text{mL}\right)=\text{OD}_{260}\times\text{Dilution factor}\times 1.1\times 10^{12} \quad (2)$$

3.4 Measurement of Viral Productivity

It is important to analyze viral productivity when a new virus is used. Good productivity of HAdV-5 can be obtained in HEK-293 cells as they express high levels of integrins and CAR receptor. However, chimeric vectors lacking the HAdV-5 fiber protein may not efficiently infect permissive HEK-293 cells, making vector amplification inefficient and viral productivity per cell very low, as it happens for chimeric HAdV-5/40 vectors [9].

$$\text{Productivity}\,(\text{vg}\,/\,\text{cell}) = \frac{\text{Titer}\,(\text{vg}\,/\,\text{mL}) \times \text{Total viral volume}\,(\text{mL})}{\text{Number of cells used}} \tag{3}$$

$$\text{Productivity}\,(\text{IU}\,/\,\text{cell}) = \frac{\text{Titer}\,(\text{IU}\,/\,\text{mL}) \times \text{Total viral volume}\,(\text{mL})}{\text{Number of cells used}} \tag{4}$$

3.5 Analysis of Adenoviral Replicative Cycle

Adenoviral replicative cycle represents the time elapsing between the adenovirus entry and the progeny release from the cell. Since a premature harvest leads to a poor production due to the insufficient virion encapsidation and maturation, it is crucial to know the viral cycle to achieve a high production yield. Similarly, in a late harvest most of the particles may have been released to the culture medium, leading to a harder purification and bigger material expenditure. Therefore, to optimize the adenovirus production is important to determine the optimal harvest time in which the viral particles are mostly mature and still are inside the cells.

Interestingly, as a consequence of the role of the fiber in the entry and trafficking of the virus particle towards the nucleus, replacement of the fiber protein is enough to alter the length of the virus life cycle [9]. For example, the cycle of the chimeric adenovirus HAdV-5/40s is delayed 20–24 h compared to the HAdV-5 cycle. Therefore, before performing a large-scale production, the harvest time for a new chimeric adenovirus should be empirically optimized to maximize the yield.

1. Seed 150,000 HEK-293 cells per well of a 24-well plate. Add 500 μL of growing medium. Samples must be harvested at 24, 36, 40, 44, 48, 52, 56, and 60 h post-infection, with a minimum of $n = 3$ wells per time point (*see* **Note 9**).
2. Add 3.75×10^6 Infection Units (IU) from your pre-stock to 12.5 mL of infection medium. Mix by inversion. These values allow to infect 25 wells at an MOI of 5 in a volume of 500 μL/well.
3. Replace the growing medium with 500 μL of the medium-virus mix generated in the previous step (*see* **Note 10**).
4. Fifteen hours later, replace the medium with fresh infection medium to remove the excess of virus used in the infection.

5. Harvest the samples at the indicated times. Collect both supernatant and cells in a tube and store it at −80 °C. Repeat for all the sampling times.
6. Freeze/thaw three times.
7. Thaw the samples and centrifuge for 5 min at 4,500 × *g*. Transfer the supernatant into a new tube and discard the pellet.
8. Plate new HEK-293 cells in a 96-well plate at a 70 % of confluence to titrate the samples (*see* **Note 11**).
9. From each sample in **step 7** save volumes of 10, 1, 0.1, or 0.01 μL and dilute to a total volume of 100 μL of infection medium.
10. Forty-eight hours after the infection count the infected cells by using the reporter gene or the hexon-antibody immunocytochemistry-based protocol. Calculate the IU/mL per sample, as described in Subheading 3.3.1. The time point with the highest value is the optimal time for harvesting the vector. As observed in Fig. 1, 56 h is the recommended harvesting time for the chimeric HAdV5-40s vector (*see* **Note 12**).

3.6 Chimeric Adenoviral Production from a Pre-stock

1. Plate twenty 150-mm petri dishes with HEK-293 cells at a confluency of 70 %. Add growing medium up to a final volume of 18 mL.
2. Twenty-four hours later, replace the growing medium with 12 mL of infection medium.
3. Calculate the amount of virus to infect twenty 150-mm plates with an MOI of 5, and dilute in 20 mL of infection medium. Add 1 mL of the mix per plate.

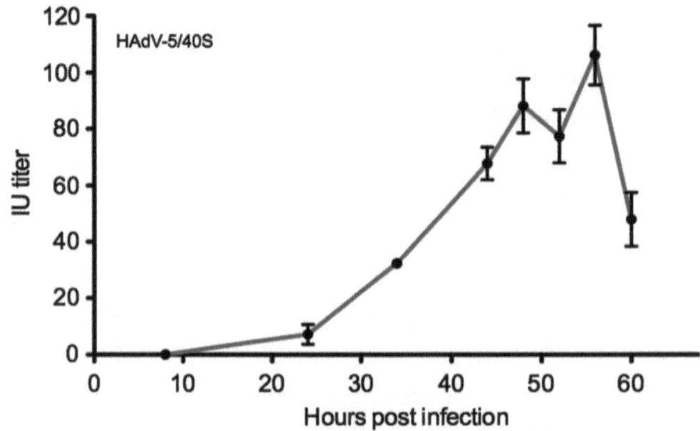

Fig. 1 Replicative cell cycle of the chimeric HAdV-5/40S vector in HEK-293 cells. According to this, the replicative cycle is between 48 and 56 h post-infection. The values are represented as percentage of the highest titer obtained

4. Optional: Add polybrene to a final concentration of 9 μg/mL to facilitate the interaction between cells and adenoviral particles. It is highly recommended when the IU/PP ratio of your pre-stock is worse than 1/100.
5. At the optimal harvest time collect cells and supernatant in 50-mL tubes.
6. Centrifuge for 5 min at 300 × *g*. Pool the pellets and resuspend it in 15-18 mL of supernatant.
7. Perform three freeze/thaw rounds (−80 °C/37 °C).
8. After the last thaw, centrifuge the cell lysate for 5 min at 4,500 × *g* and recover the supernatant.
9. Purify the virus as described in Subheading 3.2.

3.7 Production in Suspension 211BS Cell Cultures

There are some adenoviral serotypes whose productivity per cell is very inefficient [10, 11]. Therefore classical standard amplification procedures must be avoided and more efficient systems like large-scale production in cell suspension are highly recommended. This is the case of the enteric adenoviruses as the Human Adenovirus 40 serotype (HAdV-40), which due to their inefficient productivity are also known as *fastidious virus*. Interestingly, chimeric Adenovirus 5/40S (HAdV-5/40S) carrying the HAdV-40 short fiber on the HAdV-5 capsid preserves the enteric tropism of the HAdV-5/40S [12] but also the inefficient productivity.

To address this issue we have developed a protocol (*see* Fig. 2) based on the addition of polybrene during the amplification steps in combination with the use of 211BS cells (expressing constitutively HAdV-5 fiber) as producer cells [9]. First, polybrene interacts with negatively charged adenoviral capsids facilitating their interaction with the cell membrane [13, 14], which increases virus transduction. Second, 211BS cells allow to generate mosaic virions containing both F5 and F40S fiber proteins. The fiber mosaicism improves the infectivity of the chimeric virions during the amplification cycles by facilitating an efficient cell entry mediated by the CAR-F5 interaction, which together with the addition of polybrene reduces the number of amplification cycles and the duration of the process. Finally, to further facilitate production of the chimeric vectors, 211B cells are grown in suspension, thus allowing to easily up-scale the production process in bioreactors.

3.7.1 First Viral Amplification Step in 211BS from Pre-stock with Polybrene

The procedure described here has been designed to use a single amplification step in a single flask of 211BS. However, variations in the number of flasks, or the number of amplification steps can be adapted easily.

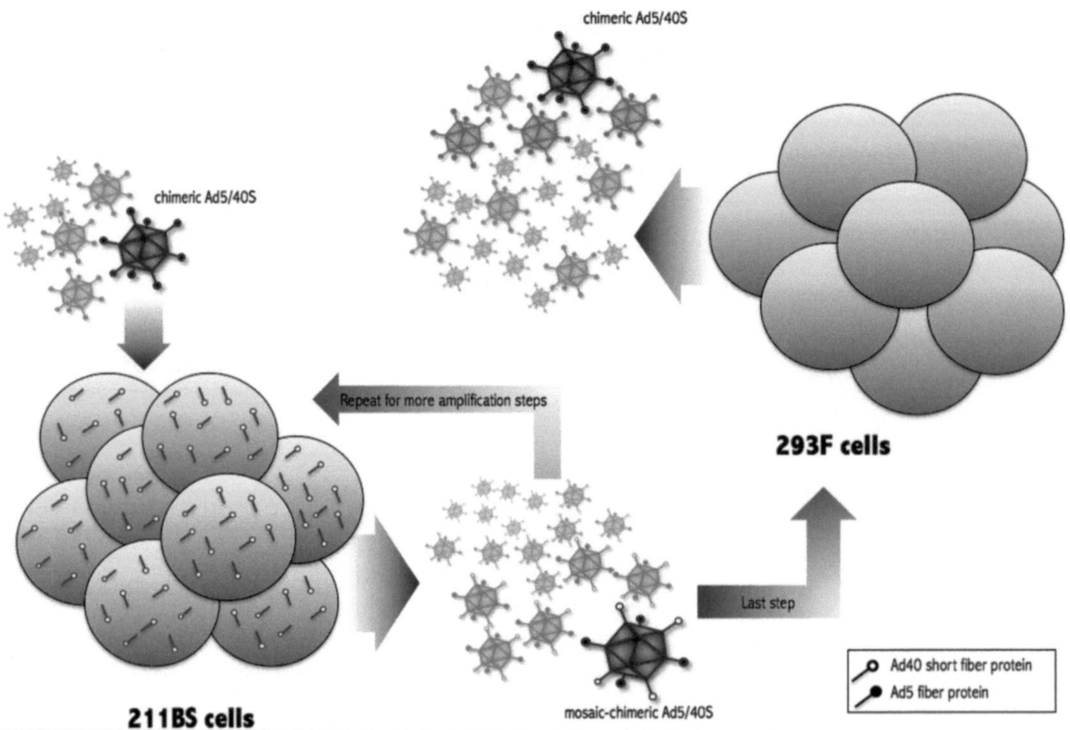

Fig. 2 Amplification strategy of chimeric HAdV-5/40S vectors. The first amplification an intermediate steps are performed in 211BS cells. Last step is carried out in 293F cells to obtain pure chimeric HAd5V-/40S vectors

1. Seed 10^6 211BS cells/mL in one 125 mL shake flask (25 mL working volume) in Infection Medium. Keep the culture in suspension by agitation in an orbital shaker at a speed of 110 rpm, 37 °C and 5 % CO_2.
2. Add polybrene to a final concentration of 9 μg/mL (*see* **Note 13**).
3. Infect cells with the vector pre-stock at an MOI of 1 (*see* **Note 14**).
4. Four hours post-infection, supplement the cell culture with 0.5 % FBS.
5. Optional step: if vector expresses a fluorescent marker protein, such as GFP, the infection efficiency can be estimated by fluorescent microscopy at 30 h post-infection (*see* **Note 15**).
6. Harvest cell cultures at 56 h post-infection and store at −80 °C.
7. Lyse cells by three freeze/thaw cycles in order to release the virus from cells.
8. Centrifuge at 1,620 × *g* for 5 min to remove cell debris.
9. Store at −80 °C.

3.7.2 Last Viral Amplification Step in 293S Cells

Last amplification step of the chimeric adenovirus should be performed in 293F cells (*see* **Note 16**).

1. Seed 10^6 293F cells/mL in 1 L shake flasks (400 mL working volume) in Infection Medium.
2. Add polybrene to a final concentration of 9 μg/mL.
3. Infect cell cultures by adding the cell lysate from **step 9** of the previous amplification procedure.
4. Change the cell culture media to fresh media at 4–6 h post-infection by centrifugation at 180 × *g* for 5 min (*see* **Note 17**).
5. Harvest cells at the optimal harvest time found in Subheading 3.5 (56 h post-infection for HAdV-5/40S) and centrifuge for 5 min at 180 × *g*.
6. Resuspend the cell pellet in 20 mL of supernatant and store at −80 °C.
7. Optional step: If the virus genome carries the Death Protein (ADP) gene, the supernatant should be concentrated down to 20 mL using a Midjet system.
8. Lyse the cell pellet by three freeze/thaw cycles. Remove cell debris by centrifugation for 5 min at 1,150 × *g*.
9. Purify the crude viral stock following Subheading 3.2.

4 Notes

1. Before digesting, check if the adenoviral sequence has internal *Pac*I cleavage sites. If this is not the case, use another restriction enzyme to digest the bacterial sequences.
2. It is recommended to use a control plate transfected with an irrelevant plasmid to test the PEI's toxicity.
3. Use a control plate to compare the cytopathic effect.
4. It is recommended to use a noninfected plate as control.
5. As freeze/thaw cycles affect the stability of the vectors, we recommend to aliquot vectors in small volumes (e.g., 10, 50, and 100 μL aliquots).
6. To calculate the titer by "end point dilution" do not count the number of infected cells because in the positive wells is possible to find a high number of infected cells if waiting for more than one replicative cycle.
7. When using a different resuspension buffer, add SDS to a final concentration of 0.1 % to disrupt the capsids.
8. The OD_{260} must be within the lineal range of your spectrophotometer. If not, repeat the previous steps with different viral dilutions.

9. The indicated time points are only a suggestion. Different time points may also be used. In addition, it is also recommended to seed three extra noninfected wells as negative controls.
10. HEK-293 cells are poorly attached to the plate surface, especially after being infected by an adenovirus. All media replacements must be done very gently.
11. Avoid working with confluences higher than 80 % because highly confluent cells are poorly infected by the adenovirus, and leads to underestimation.
12. It's recommended to discard the values when the percentage of positive cells is higher than a 10 %. In higher percentages probably some of the positive cells are infected by more than one infectious particle, which will lead to underestimation of the titer. If most of the time points have percentages of infection higher than 10 % adjust the dilutions properly from **step 9** and repeat the experiment.
13. Polybrene-mediated enhancing effects on adenovirus infection are observed only when using Freestyle serum-free medium, whereas SFMII medium completely blocks the effect of polybrene. This has also been described for other cationic molecules such as polyethilenimine (PEI) [15].
14. MOI is the number of viral infection units per cell and it depends on the cell type and the environmental conditions during infections.
15. The time at which the marker protein is visible at the fluorescent microscopy depends on the viral cycle of each vector. For example, 30 h post-infection for Ad5 or 48 h post-infection for Ad5/40.
16. HAdV-5/40S produced by 211BS cells are expected to have both, F5 and F40S proteins (mosaic–chimeric HAdV-5/40S), whereas HAdV-5/40 produced by 293F cells should only display F40S on their surface (chimeric HAdV-5/40S). In order to maximize the amplification of HAdV-5/40S, these vectors should be grown in 211BS. However, to obtain pure chimeric (not mosaic) HAdV-5/40S particles, the last step of amplification has to be performed in 293F cells.
17. Most viral particles infect the cells during the first 4–6 h post-infection. After this time, it is important to change the medium to clear the viral particles that have not entered into the cells and thus, remove the contaminating chimeric-mosaic particles used in the infection from the final step.

References

1. Kojaoghlanian T, Flomenberg P, Horwitz MS (2003) The impact of adenovirus infection on the immunocompromised host. Rev Med Virol 13:155–171
2. Haddada H, Cordier L, Perricaudet M (1995) Gene therapy using adenovirus vectors. Curr Top Microbiol Immunol 199(Pt 3):297–306
3. Wivel NA, Gao G, Wilson JM (1999) Adenovirus vectors. The development of human gene therapy. Cold Spring Harbor monograph series, pp 87–110
4. Thomas CE, Ehrhardt A, Kay MA (2003) Progress and problems with the use of viral vectors for gene therapy. Nat Rev Genet 4:346–358
5. Verma IM, Weitzman MD (2005) Gene therapy: twenty-first century medicine. Annu Rev Biochem 74:711–738
6. Gall J, Kass-Eisler A, Leinwand L, Falck-Pedersen E (1996) Adenovirus type 5 and 7 capsid chimera: fiber replacement alters receptor tropism without affecting primary immune neutralization epitopes. J Virol 70: 2116–2123
7. Nakamura T, Sato K, Hamada H (2003) Reduction of natural adenovirus tropism to the liver by both ablation of fiber-coxsackievirus and adenovirus receptor interaction and use of replaceable short fiber. J Virol 77:2512–2521
8. Roberts DM, Nanda A, Havenga MJ, Abbink P, Lynch DM, Ewald BA et al (2006) Hexon-chimaeric adenovirus serotype 5 vectors circumvent pre-existing anti-vector immunity. Nature 441:239–243
9. Miralles M, Segura MM, Puig M, Bosch A, Chillon M (2012) Efficient amplification of chimeric adenovirus 5/40S vectors carrying the short fiber protein of Ad40 in suspension cell cultures. PLoS One 7:e42073
10. Sherwood V, Burgert HG, Chen YH, Sanghera S, Katafigiotis S, Randall RE et al (2007) Improved growth of enteric adenovirus type 40 in a modified cell line that can no longer respond to interferon stimulation. J Gen Virol 88:71–76
11. Tiemessen CT, Kidd AH (1994) Adenovirus type 40 and 41 growth in vitro: host range diversity reflected by differences in patterns of DNA replication. J Virol 68:1239–1244
12. Rodriguez E, Romero C, Ferrer M, Burgueño JF, Rio A, Gil E, Hamada H, Bosch A, Gasull MA, Fernandez E, Chillon M (2006) Therapeutic potential of the Chimeric Adenovirus 5/40 as a vector for intestine directed gene therapy. Mol Ther 13:S5
13. Arcasoy SM, Latoche JD, Gondor M, Pitt BR, Pilewski JM (1997) Polycations increase the efficiency of adenovirus-mediated gene transfer to epithelial and endothelial cells in vitro. Gene Ther 4:32–38
14. Jacobsen F, Hirsch T, Mittler D, Schulte M, Lehnhardt M, Druecke D et al (2006) Polybrene improves transfection efficacy of recombinant replication-deficient adenovirus in cutaneous cells and burned skin. J Gene Med 8:138–146
15. Geisse S, Di Maiuta N, Ten Buren B, Henke M (2005) The secrets of transfection in serum-free suspension culture. In: Gòdia F, Fussenegger M (eds) Animal cell technology meets genomics. ESACT Proceedings. Springer, Netherlands, pp 373–376

Index

Miguel Chillón and Assumpció Bosch (eds.), *Adenovirus: Methods and Protocols*, Methods in Molecular Biology, vol. 1089, DOI 10.1007/978-1-62703-679-5, © Springer Science+Business Media, LLC 2014

If you have any concerns about our products
you can contact us on
ProductSafety@springernature.com

In case Publisher is established outside the EU,
the EU authorized representative is:
Springer Nature Customer Service Center GmbH
Europaplatz 3, 69115 Heidelberg, Germany

Printed by Libri Plureos GmbH
in Hamburg, Germany

MIX
Papier aus verantwortungsvollen Quellen
Paper from responsible sources
FSC® C105338

If you have any concerns about our products,
you can contact us on
ProductSafety@springernature.com

In case Publisher is established outside the EU,
the EU authorized representative is:
Springer Nature Customer Service Center GmbH
Europaplatz 3, 69115 Heidelberg, Germany

Printed by Libri Plureos GmbH
in Hamburg, Germany